The complex phenomenon of limb regeneration that occurs in some Amphibia involves unique molecular and cellular mechanisms. When a limb is amputated, a new one is produced by the transformation of the remaining adult limb tissues into an embryonic-like cell mass, called the blastema. The sequence of developmental events shown during regeneration not only reproduces the normal process of cell differentiation, but also that of pattern formation, since the regenerated limb is a faithful copy of the original. It is argued that for such a change to occur in terminally differentiated cells and for such a precise and exact morphogenetic process to occur, processes similar to those that are encountered in normal embryogenesis are reinitiated during regeneration.

The first part of the book elaborates on the mechanisms of limb regeneration and covers the regenerative powers of various different animals, the initial events of wound healing, blastema formation, the differentiation process, nerve dependence, induction, and *in vitro* models. The second part covers the mechanisms of patterning during limb regeneration and the molecular mechanisms involved. The book concludes with a short survey of the various molecular advances and techniques that are currently available to researchers working on regeneration.

This is the first book that attempts to describe and analyze the mechanisms of both limb regeneration and patterning, by incorporating the information obtained from older experiments with the many new advances in molecular and cellular biology that have occurred in recent years.

DEVELOPMENTAL AND CELL BIOLOGY SERIES

EDITORS

P. W. BARLOW   J. B. L. BARD   P. B. GREEN   D. L. KIRK

LIMB REGENERATION

Developmental and cell biology series

SERIES EDITORS

P. W. Barlow, *Long Ashton Research Station, University of Bristol*
J. B. L. Bard, *Department of Anatomy, Edinburgh University*
P. B. Green, *Department of Biology, Stanford University*
D. L. Kirk, *Department of Biology, Washington University*

The aim of the series is to present relatively short critical accounts of areas of developmental and cell biology where sufficient information has accumulated to allow a considered distillation of the subject. The fine structure of cells, embryology, morphology, physiology, genetics. biochemistry and biophysics are subjects within the scope of the series. The books are intended to interest and instruct advanced undergraduates and graduate students and to make an important contribution to teaching cell and developmental biology. At the same time, they should be of value to biologists who, while not working directly in the area of a particular volume's subject matter, wish to keep abreast of developments relevant to their particular interests.

OTHER BOOKS IN THE SERIES

R. Maksymowych *Analysis of leaf development*
L. Roberts *Cytodifferentiation in plants: xylogenesis as a model system*
P. Sengel *Morphogenesis of skin*
A. McLaren *Mammalian chimaeras*
E. Roosen-Runge *The process of spermatogenesis in animals*
F. D'Amato *Nuclear cytology in relation to development*
P. Nieuwkoop & L. Sutasurya *Primordial germ cells in the chordates*
J. Vasiliev & I. Gelfand *Neoplastic and normal cells in development*
R. Chaleff *Genetics of higher plants*
P. Nieuwkoop & L. Sutasurya *Primordial germ cells in the invertebrates*
K. Sauer *The biology of* Physarum
N. Le Douarin *The neural crest*
M. H. Kaufmann *Early mammalian development: parthenogenetic studies*
V. Y. Brodsky & I. V. Uryvaeva *Genome multiplication in growth and development*
P. Nieuwkoop, A. G. Johnen & B. Albers *The epigenetic nature of early chordate development*
V. Raghavan *Embryogenesis in Angiosperms: a developmental and experimental study*
C. J. Epstein *The consequences of chromosome imbalance: principles, mechanisms and models*
L. Saxén *Organogenesis of the kidney*
V. Raghavan *Developmental biology of the fern gametophytes*
R. Maksymowych *Analysis of growth and development in* Xanthium
B. John *Meiosis*
J. Bard *Morphogenesis: the cellular and molecular processes of developmental anatomy*
R. Wall *This side up: spatial determination in the early development of animals*
T. Sachs *Pattern formation in plant tissues*
J. M. W. Slack *From egg to embryo: regional specification in early development*
A. I. Farbman *Cell biology of olfaction*
L. G. Harrison *Kinetic theory of living pattern*
N. Satoh *Developmental biology of Ascidians*
R. Holliday *Understanding ageing*

LIMB REGENERATION

PANAGIOTIS A. TSONIS

The University of Dayton

CAMBRIDGE
UNIVERSITY PRESS

Published by the Press Syndicate of the University of Cambridge
The Pitt Building, Trumington Street, Cambridge CB2 1RP
40 West 20th Street, New York, NY 10011-4211, USA
10 Stamford Road, Oakleigh, Melbourne 3166, Australia

First published 1996

Printed in the United States of America

Library of Congress Cataloging-in-Publication Data
Tsonis, Panagiotis A.
Limb regeneration/Panagiotis A. Tsonis
p.   cm.–(Developmental and cell biology series)
Includes bibliographical references and index
ISBN 0-521-44149-8 (hardback)
1. Extremities (Anatomy)–Regeneration. I. Title. II. Series.
(DNLM: 1. Extremities–physiology. 2. Regeneration. WE 800
T882L 1996)
QP90.2.T76 1996
596´.031–dc20   95-22051
                                                            CIP

A catalog record for this book is available from the British Library

ISBN 0-521-44149-8 Hardback

Illustrations by Maria Demosthenous

Dedicated to my teachers

Goro Eguchi
Antonios Papafragas

Table of Contents

vii

Preface

The phenomenon of limb regeneration has been known for more than 220 years, and is, in my opinion, one of the most fascinating in modern science. I decided to write this book because I felt that a work like this was needed to explain to others the importance of this whole area, so that the subject could "take its place" among all the other books devoted to developmental biology. Because the subject has been studied for such a long time, there is a lot of work that is apparently old to the modern eye. Nevertheless, it is important, and it is particularly important to look at this work anew from a molecular perspective, since it sheds light on much of current research. As a student of regeneration, I hope that this book will provide a useful synthesis and contribution to the field. I am not the first author of a book on regeneration, and this book will by no means make reading the older books unnecessary. They will remain valuable not only for their historical interest, but also because many contain valuable ideas about the whole process of regeneration.

Several colleagues have contributed ideas and help, data and figures for this book, and they deserve proper acknowledgment. I am thankful to Susan Bryant, Bruce Carlson, and Roy Tassava for critical review of the manuscript and for valuable comments, and to Jeremy Brockes who suggested inclusion of Part III. Susan Bryant, Hans-Georg Simon, and Pierre Savard contributed unpublished results. I am indebted to Bruce Carlson for his hospitality in Ann Arbor and for his reprint collection, which was a valuable source of information. Thanks are also due Richard Borgens, Susan Bryant, Jeremy Brockes, Patricia Ferretti, Ronald Evans, Malcolm Maden, Marcia Newcomer, Carl Pabo, Spyros Papageorgiou, and Roy Tassava for providing figures that are included in the book.

I would also like to thank Maria Demosthenous for providing many of the illustrations.

Panagiotis A. Tsonis
Dayton, Ohio

PROLOGUE

On Myths
and Realities

Oh strangest miracles! Oh new happenings!

–ECCLESIASTIC HYMN

Prometheus shrugged

Among the vast riches of Greek mythology, the story of Prometheus is one of the most dramatic examples of punishment imposed by the gods. Against the gods' orders, Prometheus brought to the human race the fire that symbolized knowledge. The gods condemned him to exile in the far-away mountains of Caucasus, where he was bound in chains. Every day an eagle would come and eat his liver, and every night the lost liver would be regenerated. This remarkable story is not only tragic in its theme, presenting the never-ending human quest for knowledge, but it also depicts a tangible physiological process. The liver can regenerate, and life without the liver cannot be sustained beyond a single day. The regeneration of body parts has marked the paths of human imagination since ancient times. But what is most amazing in such stories is the instinct of the human race, this unmistakable sense that has driven our evolution and civilizations. The fire of knowledge brought by Prometheus symbolizes the beginning of the new human revolution, the one that permanently marked the paths of this planet with indelible footprints.

Although liver regeneration is not the subject of this book, it deserves a note when we deal with another extraordinary aspect of regeneration: that of the limbs in some amphibian species, mainly salamanders. This process is not just a simple replacement by cell proliferation as seen in replacement of removed skin or liver. It involves certain sets of cellular activities that are connected to mechanisms of development, differentiation, and possibly cancer. The basic knowledge that can be gained by the study of limb regeneration could prove of paramount importance in the fields of embryology, gene regulation, and medicine, and it is my intention to inform the reader about this potential.

From Aristotle to Spallanzani

Aristotle noticed the regeneration of the tails in lizards, a phenomenon different from limb regeneration but somewhat connected to the amphibian counterpart by virtue of parallelism of some inductive mechanisms. Aristotle apparently never studied amphibians, partly because of the loathing of the ancients for amphibians, especially salamanders. The salamander was viewed as a fearsome animal capable of killing upon contact anything alive. This exaggerated image can be seen in tales of annihilation of a whole army that passed through a river a salamander had crossed before. In other cases, the salamander was portrayed as an animal that could survive fire (this belief was still evident even in da Vinci's time). Despite all this, the spark of human curiosity eventually led scientists, including Spallanzani, to discount such myths and to experiment with the animal.

Spallanzani published his *Essay on Animal Reproductions* in 1768. The following year the essay was translated from Italian into English by Maty. Spallanzani described in great detail the regeneration of legs, tail, and jaw in the aquatic salamander. His work is spectacular for its time. One has only to realize that such developmental work came more than a century before anything of fundamental importance was known about embryo development; thus, the essay is literally one of the oldest experiments in developmental biology. Spallanzani did not flatly report what he observed. With the limited scientific means available at that time, he was able to notice differences between epimorphic and tissue regeneration, the production of heteromorphic regenerates, as well as differences in the morphology between developing and regenerating limb (see also Dinsmore, 1991).

1996

Two hundred and twenty-eight years later, scientific experimentation has come a long way. We now have tools to isolate and study the blue-prints of life, to study the function of important proteins involved in developmental events, and to correlate these proteins with diseased states. The molecular framework has been laid for many developmental processes. What about Spallanzani's developmental experiment? What do we know now that can help us understand limb regeneration? It is the main scope of this book to describe the molecular and cellular framework that has been, and can be, applied to this field. During the past 10 years, work has started at the molecular level, and exciting results are accumulating; these results could eventually throw light on the basics. In describing this new work, however, I do not intend to ignore the previous work that has marked the pages of numerous journals from the beginning of this century. At the same time, I do not wish to overwhelm the reader with extensive material that can be found in other publications. Of course, I will try to incorporate and bridge the old with the new in an attempt to put interesting old ideas into a new perspective, "brushed with the fresh paint" of modern scientific methodology. I have a second goal in mind as well: to stress the importance of this specific knowledge to other systems. I believe, and I hope I will convincingly demonstrate, that information gathered from the limb regeneration field could prove essential to the fields of gene regulation and cell differentiation of normal and abnormal states alike.

PART I

MECHANISMS OF LIMB REGENERATION

1

Capacity of Limb Regeneration in Vertebrates

1.1 A few good amphibia

How widespread is limb regeneration among vertebrates? The ability of different vertebrates to regenerate a limb has become the basis for interesting discussion and speculation. In fact, complete limb regeneration in adult animals is limited only to a few urodeles. Regeneration of limb structures, however, can occur in other amphibia, including anura, and in other vertebrates such as mice or rats. Regeneration of the fingertip has also been observed in humans. The question is, is there a principle that underlies these regenerative processes in vertebrates but has more pronounced effects in some urodeles? And if so, what is it and how could this help us improve these abilities in higher vertebrates?

The phylogenetic distribution of limb regeneration in amphibia has been treated in the past by Scadding (1977, 1981). He reported that most urodeles show excellent regenerative capacities (Table 1.1). Some exceptions include *Necturus maculosus, Siren intermedia,* and two

Amphiuma species. In these animals, limb regeneration is absent or heteromorphic. Why some urodeles are able to regenerate and others are not remains a mystery. Scadding points out that in urodeles, limb regeneration is correlated with the size of the animal; large animals are not able to regenerate or can only do so very slowly. In anuran amphibia the picture is very different. Only some species are able to regenerate as adults, and even in these cases regeneration is heteromorphic. Regeneration has been observed in some species of Discoglossidae, Pipidae, and Hylidae, but never in Bufonidae. What adds to the puzzle here is that in anura there is no correlation between size and regenerative capacity. Thus, size should be excluded if we are searching for a principle underlying limb regeneration in amphibia. Limb regeneration in anura, however, is very good in premetamorphic animals, even though it seems to happen by tissue regeneration rather than by the formation of the regeneration blastema. We could speculate from this that a mechanism tied to certain developmental processes might be very influential for regeneration. But, of course, this is not true for newts and salamanders that regenerate readily before or after metamorphosis.

What else might be so different among amphibia that can be correlated with the regenerative capacity? Since adequate innervation is very important for regeneration to take place (see Chapter 6), the quantity of innervation has been examined in urodeles and anurans. Again, as in the case of size, there is no correlation between quantity of innervation and ability to regenerate in urodeles, but in anura there seems to be a correlation (Scadding, 1982). For the sake of discussion, I would like to mention the genome size. It is true that the urodele species that cannot regenerate a limb also have the largest genomes. *Amphiuma* has from 150 to 200 pg of DNA per diploid cell, whereas urodeles with good regenerative capacities have at most about 100 pg per diploid cell (Olmo, 1973). Relative to this, it has been observed that among all Hylids the frog *Hyla versicolor* (a tetraploid frog) showed the poorest regeneration (Scadding, 1981). Scadding and Vinette (1978) have also reported that triploid *Amblystoma* larvae regenerated the limbs normally. If the size of the genome has anything to do with regenerative capacities, then molecular events (that we know nothing about) should be involved. One explanation could be that the large genome of some salamanders and newts has resulted not only in the excess "junk" DNA, but also in the duplications of structural genes and the consequent pro-

Table 1.1 Limb regeneration in adult amphibians.

	Species	Snout-vent length, mm	Limb regeneration	DNA Content pg/nucleus(diploid)
Urodela				
Ambystomatidae	Ambystoma tigrinum	90	Variable-normal, heteromorphic, or absent	57
	Ambystoma maculatum	99	Usually normal; occasionally heteromorphic	88
	Ambystoma laterale	62	or absent	102
	Ambystoma	73	Normal	---
	jeffersonianum	59	Normal	55
	Ambystoma opacum		Usually normal;	
		50-100	occasionally heteromorphic	80
	Ambystoma mexicanum		Normal	
Salamandridae	Triturus pyrrhogaster	90	Normal	---
	Triturus cristatus	90	Normal	38
	Triturus helveticus	45	Normal	---
	Triturus vulgaris	40	Normal	48
	(taeniatus)	60	Normal	45
	Triturus alpestris	90	Heteromorphic	60
	Salamandra salamandra	50	Normal	80-90
	Notophthalmus	73	Normal	---
	viridescens	75	Normal	60
	Taricha torosa			
	Taricha granulosa			
Plethodontidae	Plethodon cinereus	35	Normal	42
	Plethodon dorsalis	45	Normal	---
	Plethodon glutinosis	54	Normal	43
	Desmognathus	38	Normal	36
	ochrophaeus	62	Normal	36
	Desmognathus fuscus	39	Normal	71
	Eurycea bislineata			
Proteidae	Necturus maculosus	182	Absent	190
Sirenidae	Siren intermedia	328	Absent or heteromorphic	107
Amphiumidae	Amphiuma tridactylum	500	Heteromorphic or absent	---
	Amphiuma means	500	Heteromorphic or absent	190
Anura				
Discoglossidae	Alytes obstetricans		Absent	
	Bombina bombina		Heteromorphic	
	Bombina variegata		Heteromorphic	
	Discoglossus pictus		Heteromorphic	
Pipidae	Xenopus laevis		Heteromorphic	6
	Xenopus mulleri		Heteromorphic	---
	Hymenochirus boettgeri		Heteromorphic	---
Bufonidae	Bufo quercicus	14	Absent	---
	Bufo americanus	61	Absent	7
	Bufo marinus	110	Absent	6
	Bufo regularis		Absent	---
	Bufo andersonii		Absent	---
Hylidae	Pseudacris triseriata	24	Absent or heteromorphic	7
	Pseudacris clarki	28	Heteromorphic	---
	Hyla crucifer	23	Absent or heteromorphic	5
	Hyla squirilla	25	Absent	7
	Hyla versicolor	48	Absent	---
	Hyla cinera	38	Absent	6.5
	Hyla septentrionalis	47	Absent or heteromorphic	3
	Acris crepitans	27	Absent	---
Pelobatidae	Scaphiopus holbrooki	55	Absent	---
	Scaphiopus bombifrons	54	Absent	---
Hyperolidae	Hyperolius viridiflavus ferniquei		Heteromorphic	---
Ranidae	Rana temporaria		Absent	11
	Rana clamitans	70	Absent	11
	Rana sylvatica	50	Absent	---
	Rana palustris	60	Absent	9.5
	Rana catesbeiana	125	Absent	10
	Rana pipiens	75	Absent; rarely	10-15
	Rana esculenta	106	heteromorphic	10
	Rana ridibunda ridibunda		Absent	---
	Rana cyanophylctis	55	Heteromorphic	---
	Rana tigerina		Heteromorphic or absent	---
			Absent	

duction of more proteins. Some of these proteins exist in a window that might allow phenomena that took place during embryogenesis to be repeated with the right stimulus. Of course, all of this is hypothetical, and all I do is entertain a possibility. We do not know much about genome organization and regulation in amphibia. Existence of more copies of structural genes has not been well documented, even though it seems that the genome size correlates with cell size in urodeles; but it needs to be elucidated whether this means more proteins.

Let us elaborate some more on the amount of DNA. The presence of junk DNA may correlate with long introns that could account for the large amount of DNA. This might suggest that the exons or the coding sequences are not in multiple copies. Some limited data exist in favor of both ideas, that is large introns vs. gene duplication. First, it has been shown that the introns in the myosin gene of the newt are seven times larger than the mammalian counterpart (Casimir et al., 1988a). The large size of introns could, in fact, account for the large genome in newts. However, introns that separate newt Hox genes seem to be of the same size as those in mammals (Belleville et al., 1992). The same seems to be the case for beta-actin introns (Burns et al., 1994). These observations indicate that the intron size alone should not be considered the sole contributor to the huge amphibian genome, which in turn leaves the door open for the idea of duplication of structural genes (Del Rio-Tsonis and Tsonis, 1994).

The potential for regeneration in the few lucky urodeles is very good throughout their adult life. Moreover, it seems that an individual can repeatedly regenerate its limbs after subsequent amputations. It also seems that regeneration does not become impaired or abnormal due to repeated amputations. Spallanzani reported that the animal was able to regenerate up to six times after repeated amputation. In 1916, Zeleny published a detailed study on factors controlling the rate of regeneration. He found virtually no effect of repeated amputation (up to four times) on the rate of regeneration or the length of the regenerates. He postulated that the decrease in ability could instead be the result of age rather than the number of amputations. Kollros (1984) also reported that anuran tadpoles were able to regenerate successfully even after six sequential amputations. In fact, Kollros stated that the repeated amputations prolonged the regenerative ability of the frogs. A report by Dearlove and Dresden (1976) suggested that repeated amputations would result in abnormal regeneration. However, when newts with abnormal limbs were collected from the field, or those with induced abnormal limbs were reamputated proximal to the abnormality, regeneration proceeded normally. In fact, these animals corrected the distal abnormalities (Tsonis and Eguchi, 1985). Polezhaev (1972) has reported that repeated amputations were, in fact, able to prolong the ability of *Rana temporaria* to regenerate the limbs.

Lizards exhibit a spectacular reptilian ability to regenerate their tails by autotomy. This is a different mechanism from the one we have encountered for the amphibia. Through autotomy, the tail is aborted at a specific region when an animal has been caught. It is a mechanism for helping the animal escape from predators. Despite this ability for tail regeneration, however, lizards do not regenerate their limbs. Upon amputation, a blastema is formed but it fails to increase in size, and finally most of the cells are differentiated to cartilage. Similarly, amputation of embryonic lizard limbs resulted in no regeneration (Bellairs and Bryant, 1968). Singer (1961), however, was able to induce regeneration of limbs in the lizard *Anolis carolinensis* by augmenting the nerve supply of the amputated limb. The additional nerves were derived from the unamputated contralateral limb. This mode of wound healing might be an important element in the regeneration of the lizard limbs. As opposed to the tail, the limb tissues are packed loosely, and this leads to the shrinkage and other wound-healing characteristics of mammalian skin wounds (Barber, 1944). Such a claim, however, was not supported in the study by Bellairs and Bryant (1968).

1.2 Higher vertebrates

The picture concerning regeneration in higher vertebrates is very different. It has been observed that in humans, especially in young children, regrowth of amputated fingertips can occur. This depends on the level of amputation and on the treatment. Specifically, regeneration will be perfect within 3–4 months if the amputation happens distal to the last interphalangeal joint and if no operation is performed to close the wound with skin, but instead it is cleaned well and the dressings changed frequently (Illingworth, 1974). This is similar to the regeneration of anuran limbs that basically have lost the regenerative capacity

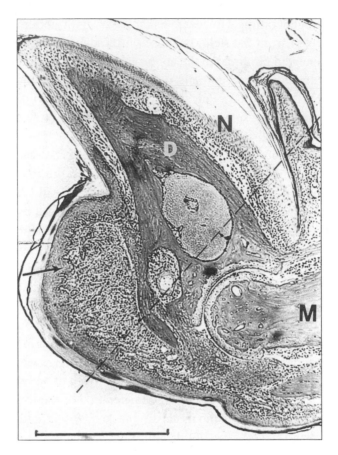

Figure 1.1 A longitudinal section of an intact mouse foretoe. The nail and the nail-bed germinal epithelium (N) insert at the level of the joint between the distal phalange (D) and the middle phalange (M). The arrow indicates invagination of dermal papillae into the fatpad. The hatched line represents amputation level 1, while amputation level 2 would pass through the extreme right margin of the photomicrograph. Bar 0.5 mm. (Courtesy R. B. Borgens.)

along the proximal–distal axis. Another similarity between amphibian and human limb regeneration is the fact that both are inhibited if a skin flap is grafted over the wound.

Similarly, mice can regrow the tips of their foretoes faithfully, depending on the amputation level, as in humans (Borgens, 1982) (Figures 1.1 and 1.2). Regeneration of the limb buds has also been observed in developing mouse limbs. Amputations were performed in developing mouse limbs *ex utero*. The stage was 7/8, which corresponds

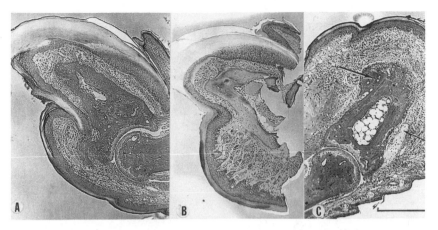

Figure 1.2 Histological sections of amputated mouse foretoe tips. A: A regenerated tip after level 1 amputation, 4 weeks, 6 days after amputation. Note the irregular outline of the distal phalange and the lack of dermal papillae in the fatpad. B: A section from the removed foretoe tips shown in A. A rudiment of the nail bed was probably left with the stump. C: A healed stump in response to amputation level 2. Note the callus formation (heavy arrow) and the retracted tendon (light arrow). Dense connective tissue is present between the severed bone and the skin. Bar 0.5 mm. (Courtesy R. B. Borgens.)

to chick 27/28. At that stage the limbs are well formed, with cartilaginous digits. Amputation performed within an individual digit primordium resulted in regeneration of the digit (Wanek, Muneoka, and Bryant, 1989) (Figure 1.3). It has also been shown that younger limb buds (of stage 1) and rat limb buds (of stage 2/3) that are amputated and cultured *in vitro* can form an apical ectodermal ridge and also occasionally produce a limb of normal appearance (Deuchar, 1976; Chan, Lee, and Tam, 1991). Polezhaev (1972) amputated adult mouse limbs at the level of the wrist and injected the animals with extracts of vitreous body (a rich source of growth factors). He obtained a generation of toelike appendages. The basis of the formation of these appendages was the remaining bone. Cartilage, connective tissue, and cornified nails grew. Experiments with rats also have shown some regeneration. Nicholas (1926) amputated the limbs of rat embryos at different days of gestation. He found that after the 14th day there was no sign of regeneration. The limb structures had been formed by the 12th day. Stimulation of limb regeneration in 2–7-day-old rats has been reported by Kudokotsev (cited by Polezhaev, 1972) by injection with cartilage extracts and by trypsin

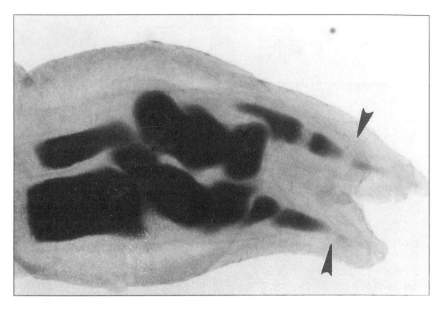

Figure 1.3 Limb regeneration in the embryonic mouse limb. The limb (stage 7/8) was amputated between the metatarsals and the tarsals. The regenerated portions of digits 1 and 5 distally to the arrowheads are shown. (Courtesy Dr. S. V. Bryant.)

and $CaCl_2$ treatment. Scharf (1961, 1963) amputated two toes of 2-day-old rats and treated the wounds with trypsin and $CaCl_2$. He observed no scab formation and the regeneration of two short toes with later appearance of nails. Scharf believed that destruction of tissues was responsible for the observed regeneration, possibly by inducing dedifferentiation. Similar claims have been presented by Rose (1942, 1945) and Neufeld (1980) using NaCl treatment in frogs and mice, respectively.

Mizell experimented with the newborn opossum and found that the nerve supply might induce regeneration. Mizell (1968) and Mizell and Isaacs (1970) transplanted supplemental nerves to the hindlimbs, which were subsequently amputated above the ankle. Regeneration of the limb resulted. Fleming and Tassava (1981), however, questioned these findings. They claimed that amputation of the hindlimb on a 4-day–post partum opossum does not transect the tibia and fibula, because the rudiments of these bones are still inside the body. Only the tarsals or the phalangeal rudiments are transected. Therefore, the limb stumps were left with distal skeletal elements that subsequently underwent embryonic development, giving the erroneous impression of regeneration.

On the other hand, birds have shown no regeneration of their limbs at any stage of development (Maden, 1981a). This inability of birds to regenerate, however, could be related to the inhibition of apical ectodermal ridge (AER) formation after the amputation. In fact, when chick limbs are amputated and covered with an ectodermal jacket that includes an AER, distal regeneration can be achieved (Rubin and Saunders, 1972). When the distal mesenchyme is removed, but the AER is intact, formation of the distal skeletal elements is permitted (Hayamizu et al., 1994). If a comparison is to be made, it seems that the regenerative abilities of mammals are more similar to those of anurans than to those of birds or urodeles.

What does all this tell us? Is there any fundamental process that can be attributed to successful limb regeneration? It is true that regeneration is more readily achieved in embryonic limbs (or young limbs) and that proper epithelial–mesenchymal interactions are very vital. And it is accepted that the regeneration capacity for more proximal structures is lost as we climb the evolutionary ladder to higher vertebrates. This in turn could mean that the regenerative capacity might exist in higher vertebrates, including humans, if the appropriate signals could be deciphered. But the phylogenetics of limb regeneration do not really provide a strong clue as to why regeneration happens in some adult urodeles. It is possible that regeneration involves specific mechanisms unique to the individual, and, therefore, its loss is linked to these unique mechanisms. For example, the axolotl possesses remarkable abilities to regenerate its tail and limbs, but it is not able, like newts, to regenerate the eye lens, another regenerative phenomenon restricted to some urodeles. In this case, it could be that limb regeneration started as a basic mechanism that duplicated limb development, but it slowly became modified to best suit the individual species. Such an idea could imply that the molecular program for regeneration is available but not accessible in higher vertebrates. Support for this comes from experiments showing that the developing limb bud and the regenerating blastema do share morphogenetic similarities, but we now know that several genes do not have the same expression patterns in the regenerating and the developing limb (see Chapter 19).

1.3 Before amphibia

Let us elaborate further on the developmental and evolutionary aspects of limb regeneration. First, we will consider the evolutionary events that led to the tetrapod amphibians. Amphibia evolved from fish. The limbs of the first amphibian tetrapods evolved by the transformation of the pelvic fins to hindlimbs, and of the pectoral fins to forelimbs. The pelvic and pectoral fins were developed by one fin fold. Fossil evidence suggests that the earliest tetrapod was an animal with leglike hindlimbs and fishlike pectoral appendages. It has been hypothesized that a subsequent homeotic transformation took place in a particular lineage, which led to pectoral fins' expressing the same homeotic genes as the pelvic appendage. These appendages then evolved to limbs (Tabin and Laufer, 1993). Comparison of Hox gene expression between fin development in zebrafish and tetrapod vertebrates has shown interesting differences. In the early budding phase, the expression of HoxA, HoxD, and *Shh* (see Chapter 18) is the same in fish and tetrapods. But in later stages, expression of HoxD 11, HoxD 12, or HoxD 13 was not observed in the cells of the anterior fin, but was observed in cells of the limb. The results suggest that regulation of the expression of Hox genes might account for the fin-to-limb transition (Sordino, van der Hoeven, and Duboule, 1995). Homeotic transformations are known to involve genes containing a DNA-binding region, the homeo box. Examples can be found in the *Drosophila* mutation *Antennapedia*, where a mutation has produced a limb instead of an antenna. Other examples of homeotic transformations come from experiments where animals have been manipulated by Hox genes (Morgan et al., 1992). Such a hypothesis is supported by the fact that in tetrapods, homeotic genes of the HoxA and HoxD clusters (Chapter 18) show almost identical expression patterns in the fore- and hindlimb.

If limb regeneration is an existing phenomenon that was modified during evolution, we would expect that the equivalent organs of the amphibian ancestors would possess regenerative capacities. In fact, this is true; some fishes do regenerate their fins, the structures that were replaced by limbs when the amphibia evolved (Wagner and Misof, 1992). Regenerative capacity, therefore, was not lost in all species during these evolutionary events. Does this imply a possible role of Hox genes in regeneration? This will be analyzed in detail later (Chapter 18), but it is worth mentioning here an interesting finding: Amputated fins respond to retinoic acid the same way amphibian limbs do; regeneration is proximalized (Geraudie et al., 1993). All these results suggest that the capacity for regeneration might have certain common mechanisms in different animals and that these mechanisms were preserved during evolution. This in turn implies that limb regeneration is not adaptive (Goss, 1992) but has a developmental and evolutionary origin, and its gradual absence from higher animals could be the effect of certain developmental and evolutionary modifications.

This book will concentrate on details of the limb regeneration process in amphibians. These animals have provided the best material for the study of this phenomenon. When appropriate, comparison with knowledge from other systems will be included and discussed.

1.4 Addendum: Regeneration in arthropods

The earliest discovery of regeneration of legs was in fact reported in Crustacea by Rene-Antoine Reaumur in 1712. Subsequently, such regeneration was noted for insects. This type of limb regeneration, however, is different from the one we encounter in vertebrates. Arthropods have to molt in order to grow (Goss, 1969). The reason for this is the hard exoskeleton that covers the body. The enzymes responsible for degrading the exoskeletal proteins are secreted from the integumentary tissues that underlie the exoskeleton. These enzymes are acidic and basic proteinases (O'Brien and Skinner, 1988). Protein synthesis during the different stages of molting is regulated by the molt cycle (Stringfellow and Skinner, 1988). Therefore, regeneration of the legs depends on this ability of the animal to molt. In other words, when a leg is severed, the regenerated limb will be present after the next molting (Goss, 1969; Skinner and Graham, 1970). When an appendage is amputated by autotomy or amputation along its length, it regenerates inside the next proximal segment (Figure 1.4). The room in the proximal segment is provided by the muscle. It is not clear whether the degenerated muscle provides dedifferentiated cells that contribute to the regenerate. The association between molting and regeneration is in fact two-way: It has been demonstrated that regeneration will promote molting (Skinner and Graham, 1970). The central nervous system does not seem to play a decisive role in limb regeneration in arthropods (Goss,

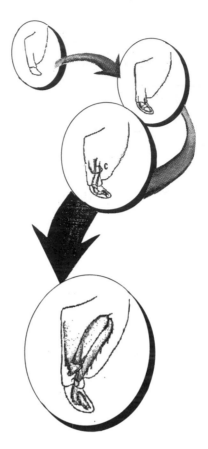

Figure 1.4 Regeneration of insect limb inside the next proximal segment. After autotomy the new leg occupies the space in the coxa due to muscle degeneration. Two weeks after autotomy, regeneration is complete. The new limb is folded in the coxa (c) and will be extended at the next molt. (Adapted from R. J. Goss, 1969.)

1969). Rather, the hormone ecdysone that controls ecdysis (the loss of the exoskeleton) should be implicated in the stimulation of regeneration. On the other hand, acceleration of limb regeneration has been seen after removal of the eyestalk, which is the source of chromatophorotropic hormones and the so-called X-organs, which secrete an antimolting hormone (Skinner and Cook, 1991).

2

The Amputation–
The Early Events

It is logical to assume that the proper amputation (that is, the removal of the whole part of a limb, and not just an injury) provides the animal (in our case, the salamander) with the necessary signal for regeneration to start. In this respect, the signal must be very specific to initiate all the developmental phenomena, and irrelevant to simple tissue regeneration observed during injuries. It is of interest, therefore, to dissect the very early responses – the first minutes or hours – in order to decipher these signals. Some insights have come from some experiments performed by Thornton (1949) dealing with the effects of beryllium on limb regeneration. Thornton found that this substance could inhibit regeneration if the limbs were treated immediately after amputation. Specifically, after amputation the animals were immersed in beryllium solution for about 2 minutes, and then they were removed and washed extensively with water. The treated limb showed inhibition of regeneration. No blastema was formed, the wound epithelium was thick, and extensive extracellular matrix was accumulated beneath it (Figure 2.1).

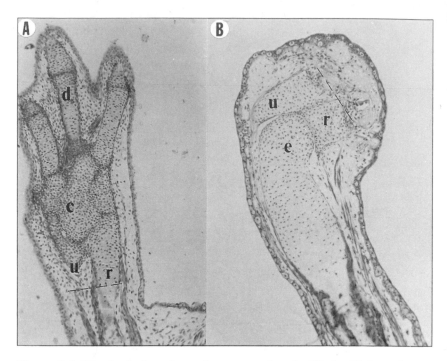

Figure 2.1 Histological sections of regenerated axolotl limbs 30 days after amputation. A: A control limb regenerated normally with digits. B: A limb treated with beryllium. No regeneration can be seen. The epithelium has covered the cone, and extensive extracellular matrix can be observed beneath the wound epithelium, with no apparent blastema formation. The dashed line represents the level of amputation. e: elbow, u: ulna, r: radius, d: digit.

If treatment with beryllium occurs 10 minutes after amputation or if amputation happens after washing off the beryllium, regeneration is not affected. Interestingly, in larvae (3–4.5 cm in length) there were differences in the requirement of beryllium along the proximal–distal axis. For regeneration to be inhibited following amputation through the arm, a treatment lasting three times longer was necessary. For younger larvae (1.6–1.8 cm in length) such a condition did not apply. Such experiments indicated that signals that might initiate the process of regeneration are very rapid and, if blocked, regeneration does not take place. Such rapid signals are reminiscent of the ones involved in signal transduction mediated by second messengers. Other studies have correlated the presence of second messengers, such as inositol trisphosphates (IPs), in the early events following amputation (Tsonis, English, and Mescher, 1991). It was found that IPs were dramatically increased during the first minute after amputation. Interestingly, however, the levels of IPs were decreased below the normal level when the amputated limbs were treated with beryllium (Figure 2.2). Lithium is also known to decrease levels of IPs in different systems and is known to act through the nervous system. The effects of lithium on regeneration were very similar to those of beryllium (Tsonis, English, and Mescher, 1991). We begin now to see

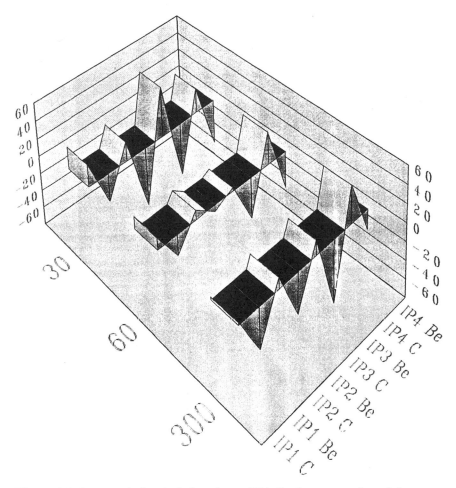

Figure 2.2 Increase in inositol phosphates (IP1-4) after amputation of the axolotl larvae limbs. Treatment with beryllium (Be) decreases these levels when compared to the untreated (C). Zero level is the level of inositol phosphates in normal unamputated limb. 30, 60, and 300 represent post-amputation time in seconds.

a clearer picture of what might be happening after amputation. Once the limb has been removed, a signal entangling second messengers is transmitted through the nerves or perhaps through other cell types, even though there is no proof for such a mechanism yet. This in turn initiates the synthesis of necessary processes for cell dedifferentiation and subsequent regeneration.

It is true enough that changes in protein synthesis occur soon after amputation. Suppression and induction of proteins happen within 3 hours post-amputation. Loss of proteins is more prominent in the stump, whereas induction of new proteins takes place along the remaining limb and even the unamputated contralateral limb, but not in other tissues of the body (Tam, Vethamany-Globus, and Globus, 1992). The identity of these proteins is not known. These results probably indicate localized effects of amputation. It is not known the extent to which the nerves are responsible for these changes; however, the induction of the proteins in the contralateral limb argues for signal transmission through spinal interneurons. In different studies involving limbs 3 to 10 days after amputation, it was found that protein synthesis was stimulated in the brachial nerve ganglia of the amputated limb. No synthesis of new proteins was detected, but a group of basic proteins was predominantly expressed (Bao, Singer, and Ilan, 1986).

3

The Beginning of Regeneration – Wound Epithelium

The first morphologically detectable event in regeneration is the covering of the wound by a specialized epithelium, namely, the wound epithelium (WE). This happens rather fast, within hours or days depending on the size and age of the animal. This covering and the establishment of the wound epithelium are critical for regeneration. If the wound epithelium is removed, regeneration cannot proceed (Thornton, 1957). It is now believed that this specialized epithelium provides the necessary signals for dedifferentiation to the underlying stump tissue and the signals for growth to the blastema cells. This idea is supported by the finding that transplantation of the epidermal cap in urodeles can produce accessory limb production (Thornton and Thornton, 1965).

The wound epithelium is produced by migration of the epidermal cells at the edge of the amputation surface. However, the wound epithelium has its own identity; it is biochemically distinct from its progenitor, the normal epidermis. The cells of the epithelium cover the wound almost immediately after amputation. Once the cells have covered the wound, they begin to synthesize molecules specific for the establishment of the basement membrane, such as laminin and collagen type IV (Figure 3.1) (Del Rio-Tsonis, Washabaugh, and Tsonis, 1992).

As early as 1 hour after amputation, the epidermal cells show signs of migration. These signs include changes in cell shape, disappearance of hemidesmosomes, and detachment of the basal epidermal cells from the basal lamina (Norman and Schmidt, 1967; Repesh and Oberpriller, 1978, 1980). Within 2 hours after amputation, the evidence of migration is more dramatic. The migrating cells have begun to flatten as they form a sheet that adheres to and migrates across the fibrin substratum. This substratum is provided by the fibrin clot over the cut surface with the presence of erythrocytes and neutrophils (Figure 3.2). The wound surface can close within 9 to 13 hours. During the migration period there is no mitosis or DNA synthesis.

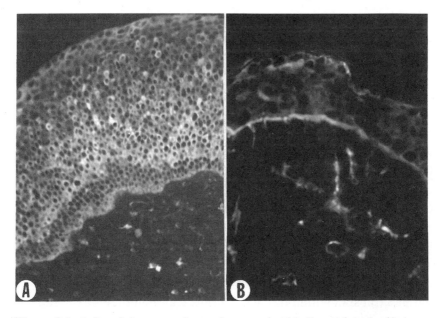

Figure 3.1 A: Laminin expression at the wound epithelium of axolotl 2 days after amputation. Note the high expression in cells at the basal layers where the basement membrane is built. Laminin is mostly absent in the underlying mesenchyme. B: Laminin expression in the blastema of the newt 4 weeks after amputation. Expression is most pronounced in the basement membrane. Some cellular structures in the mesenchyme also show expression.

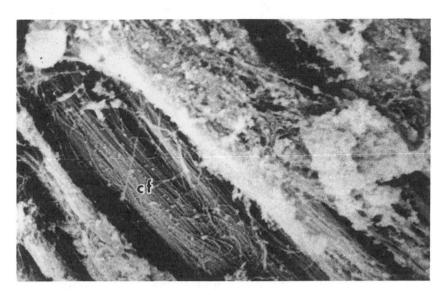

Figure 3.2 Scanning electron micrograph of the wound epithelial cells coverage of the stump and their adherence on fibers (cf) provided by the clot.

The cellular mechanism of migration has been the subject of debate. Lash (1955) believed that the cells migrate as a sheet and that the cells are loosely adherent. In contrast, Gibbins (1978) proposed that a cell can advance by passing over another cell that has already made contact with the substratum. The passing cell will cover the cell beneath it and extend its leading lamella to contact the substratum. An important point here is the availability of the extracellular matrix. The cells need this cushion in order to move. In this sense, specific gene expression events must be initiated for the matrix to be synthesized. It is true that gene expression can be modified in a cell type depending on the surrounding matrix available. For example, epithelial cells have been shown to lose their polarity and express characteristics of migrating mesenchymal cells when they are placed in collagen gels (Greenberg and Hay, 1982). Migration of the epidermal cells can be inhibited or delayed by treatment with carcinogens. During the inhibitory period there is no production of extracellular matrix to provide the substratum for the migration (Tsonis and Eguchi, 1983). As the cells migrate to cover the wound, they eventually meet. When this happens, they pile up temporarily. Once formed, the wound epithelium becomes well innervated. However, it seems that epidermal innervation is not essential for regeneration. These nerves do not permit limb regeneration in the same way that nerves in the blastema are instrumental for regeneration (Thornton, 1960; Singer and Inoue, 1964).

One of the most prominent activities of the migrating cells is phago-cytosis. The epithelial cells covering the wound can incorporate small masses of damaged cells or pieces of fibrin. When damaged tissues or foreign substances are placed underneath the epidermis, they can be sur-rounded by sheets of epithelium and removed from the body (Singer and Salpeter, 1961). Another function that has been assigned to the wound epithelium is that of histolysis of the underlying mesenchyme (Taban, 1955; Polezhaev, 1972). The inflammation caused by the amputation first brings erythrocytes and neutrophils to the area, then macrophages and mast cells, and later osteoclasts.

Goss (1956a) also demonstrated the importance of the formation of the wound epithelium. He amputated limbs and then placed them epi-dermis-free into the body cavity. No wound epithelium was formed and no regeneration resulted. Similarly, skin grafts over the amputation sur-face inhibit regeneration (Mescher, 1976). Three roles have now been attributed to the wound epithelium. First is the induction of dedifferen-tiation by histolysis of the mesenchyme. The second role is keeping the dedifferentiated cells cycling (Tassava and Mescher, 1975; Tassava and Lloyd, 1977). A third role of the wound epithelium, that of providing positional values for the start of dedifferentiation and pattern formation, has been forwarded by several scientists. Based on this view, the wound epithelium possesses a positional value that will create discontinuities once confronting a different value of the cut surface. This could initiate regeneration (see Chapter 16).

3.1 Markers

As mentioned previously, the wound epithelium synthesizes factors that are specific to this tissue. Several antibodies have been produced that cross-react with antigens restricted to wound epithelium. The WE3 anti-gen is up-regulated in the newt wound epithelium just prior to blastema formation and remains abundant until differentiation occurs (Tassava, Johnson-Wint, and Gross, 1986) (Figure 3.3). In the normal epidermis

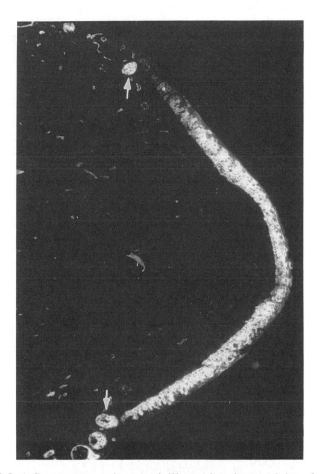

Figure 3.3 A fluorescence micrograph illustrating the reactivity of monoclonal antibody WE3 to the wound epithelium of a mid-bud regenerate of an adult newt. Note that reactivity is restricted to the wound epithelium. The outer wound epithelial cells and the distal epidermis are not reactive. The amputation level is indicated by the most distal integumentary glands (arrows). Glands and various other secretory cell types of the body are always reactive to this antibody. (Courtesy Dr. R. A. Tassava.)

only a few small round cells react to this antibody. It is now believed that this antibody reacts with an antigen related to secretion, because it is normally found in tissues specializing in transport, such as skin glands. The relation of WE3 to transport is further supported by its colocalization with the enzyme carbonic anhydrase, which is normally associated with transport. The WE3 antigen is a probable glycoprotein that elutes during gel filtration as a 660 KDa complex (Goldhamer, Tomlinson, and Tassava, 1989). A different antibody, WE4 (Tassava et al., 1993), reacts with the wound epithelium in *Notophthalmus viridescens,* whereas it cross-reacts with both epidermis and wound epithelium in the European newt *Pleurodeles waltl.* Evidence suggests that both WE3 and WE4 are actin-binding proteins. Treatment of sections or extracts with DNase I (which binds actin) dramatically reduces reactivity of both mAbs WE3 and WE4 (Tassava et al., 1993).

A third antibody, WE6, identifies still another antigen that is up-regulated in wound epithelial cells present in glandular tissues (Figure 3.4)

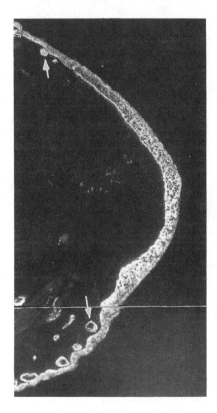

Figure 3.4 A fluorescence micrograph illustrating the reactivity of monoclonal antibody WE6 to the wound epithelium of an early bud regenerate of an adult newt. Note that the distal epidermal cells also up-regulate the WE6 antigen. The amputation level is indicated by the most distal integumentary glands (arrows). Glands and various other secretory cell types of the body are always reactive to mAb WE6. (Courtesy Dr. R. A. Tassava.)

throughout vertebrates (Estrada et al., 1993). Identification of cDNA clones reacting to this antibody in an expression library indicates that the WE6 antigen is a keratin with a molecular weight of 39 KDa when determined under reducing conditions. Finally, mAb 9G1 cross-reacts with an intracellular antigen present in the most distal part of the wound epithelium, the apical epithelial cap, and with the mesenchymal blastema cells. The expression of this antigen is required by the underlying mesenchyme and the nerve supply (Onda and Tassava, 1991).

3.2 Growth factors in the wound epithelium

The idea that the wound epithelium supplies mitotic or growth-promoting signals has been investigated. When blastema cells were grown in culture, it was found that the highest enhancement of proliferation was obtained when they were grown in the presence of extracts from the epidermal cap isolated from a 14-day-old blastema (Boilly and Albert, 1990). One factor that seems to be important for this function is the acidic fibroblast growth factor (aFGF). Members of the FGF family have in the past been found to be mitogenic in many cell types, including cultured blastema explants or blastema cells (Mescher and Gospodarowicz, 1979; Mescher, 1983). FGF has been found to be present in the wound epithelium and the mesenchyme (Hondermarck and Boilly, 1990; Boilly et al., 1991). The epidermal cap contains more aFGF than the mesenchyme. Cell membranes contain both high- and low-affinity binding sites, whereas the mesenchyme has high-affinity receptors. This suggests that the low-affinity binding sites in the epidermal cap might correlate with the high amount of heparan sulfate there, and that these sites are thus utilized as storage tissue. The binding of FGF is heparinase sensitive, which implies an association of FGF with heparin-like structures. Taking into consideration the proliferative actions of FGF on the blastema, it seems that aFGF is synthesized by the wound epithelium, but its target is the mesenchyme. Dissociation of FGF from the heparin structures might be crucial for the initiation of the dedifferentiation and blastema formation. High salt treatment can dissociate this bonding, and NaCl has been shown to induce partial blastema formation in amputated mouse limbs (Neufeld, 1980) and in frogs (Rose, 1942, 1945).

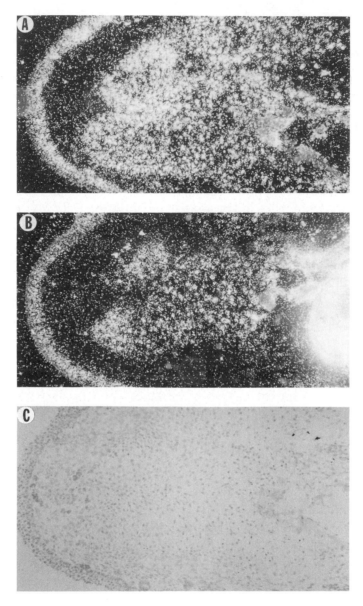

Figure 3.5 Light- and dark-field micrographs illustrating hybridization of fibroblast growth factor receptor 1 and 2 to sections of newt early digit stage regenerates. Antisense riboprobes of FGFR2 (A) and FGFR1 (B) were hybridized. An adjacent section stained with hematoxylin and eosin illustrates the histological appearance (C). At this stage of regeneration, two domains of hybridization are visible with each probe, the basal layers of the wound epithelium and the condensing cartilage of the newly forming skeletal elements. (Courtesy M. Poulin and I.-M. Chiu.)

The effects of FGF are mediated by receptors. Two such receptors, FGFR1 and FGFR2, have been cloned from the newt *Notophthalmus viridescens* (Poulin et al., 1993). Before the accumulation of blastema cells, FGFR2 is expressed in the basal layer of the wound epithelium and in the periosteum. In fact, the *bek* variant of FGFR2 (which differs from the other variant KGFR in the second half of the last Ig-like domain) is restricted in the periosteum during pre-blastema stage, while the KGFR is expressed in the basal layer of the wound epithelium. As regeneration progresses, the *bek* transcripts are present in the mesenchymal cells while the KGFR ones are present in the basal layer and the mesenchymal cells (Poulin and Chiu, 1995). These patterns are very similar to the ones observed during mouse limb development (Orr-Urtreger et al., 1993). As blastema formation progresses, expression is also seen in the mesenchyme (Figure 3.5). In other systems, FGFR2 has been involved in the epithelial–mesenchymal conversion mediated by the *fos* oncoprotein (Reichmann et al., 1992; Scotet et al., 1995). Such events are not known for the wound epithelium, but an analogous study would be very interesting. FGFR1 is expressed throughout the mesenchyme of the early to mid-bud blastema, but it is absent from the wound epithelium. During differentiation, FGFR2 is expressed in the condensing cartilage and the perichondrium. These studies suggest distinct roles of FGF receptors in limb regeneration. FGFR2 expression by the wound epithelium and mesenchyme suggests both cell types are targets of FGF. At least one source of FGF might be the wound epithelium. Support for a signaling role of FGF has been provided from studies with the developing chick limb bud. If the apical ectodermal ridge (AER) is removed, development of the limb is stalled. However, if FGF4 or FGF2 is applied after removal of the AER, development of the distal skeletal elements is restored (Niswander et al., 1993; Fallon et al., 1994). In this respect, the AER and the wound epithelium might share functional similarities in inducing the underlying mesenchyme. From these studies, it is reasonable to suggest that epidermal cells must undergo considerable biochemical changes to become a functional wound epithelium.

4

Dedifferentiation and Origin of the Blastema

The wound healing stage (0 to 5 days after amputation) is the first phase of regeneration. It is now believed that, during this phase, the necessary signals are provided to the underlying stump tissues to dedifferentiate and subsequently to form the blastema.

Dedifferentiation seems to be the key element for successful limb regeneration seen in amphibians. What actually happens is that cells of the stump tissues, including muscle, bone, nerve sheath, and connective tissues, lose the characteristics of their origin, virtually "melt down" to mononucleated, undifferentiated cells, initiate cell cycling, and produce the blastema. Fritsch (1911) showed that regeneration does not take place by direct outgrowth but by the production of undifferentiated blastema cells. Since the beginning of the century, when the term *dedifferentiation* was introduced by Driesch (1902) and Schultz (1907) and applied to the phenomena of limb regeneration by Butler (1933), there has been much discussion and experimentation on the validity of the term and of the phenomenon itself in limb regeneration. Since then, many different views have been presented of how the blastema is produced. Ideas about a blood origin and epidermal origin of the blastema were quickly and convincingly discarded. The idea of reserve or stem cells has also received some support, but the aspects of dedifferentiation have overwhelmingly prevailed.

4.1 Local origin of blastema cells

If blastema cells originate by dedifferentiation of the stump tissues, then the origin of the blastema cells should be regarded as local. In contrast, if stem cells were responsible for the creation of the blastema, they should be found throughout the limb or the body. An experiment by Butler and O'Brien (1942) gave the first serious victory to the dedifferentiation view. A whole animal (*Eurycea bislineata*) was shielded except the knee and was subsequently irradiated with X-rays at a dose that would inhibit limb regeneration. When the animal was amputated through the protected region of the limb, regeneration took place, but when the amputation was done through an exposed portion, no regeneration was obtained. This experiment strongly suggests that the blastema cells are derived from the tissues just proximal to the amputation site and therefore originate locally, and it argues against stem cells, which should be everywhere in the limb, or the body.

Chalkley (1954, 1956) realized that a paramount effect of the amputation was cell proliferation as cells undergo dedifferentiation. By counting mitoses he found that more than 80 percent of the blastema cells are provided from the connective tissues. Chalkley found that the tissue distribution of mitoses of the total blastema and connective tissue at 7 days, compared with their distribution at 37 days, follows the increase in cell numbers. Correlations were also shown between distribution of the total subepidermal mitoses and the distribution of subepidermal cells in the unamputated limb. Similarly, he showed a correlation between the number of subepidermal cells in various tissue categories that increase and the distribution of mitoses in similar or identical tissues. Subsequent studies by Muneoka, Fox, and Bryant (1986) have shown (by virtue of triploid/diploid cell marker) that dermal cells contribute 43 percent of the blastema cell population, whereas cells from skeletal tissues contribute only 2 percent. The authors suggested that the fibroblasts present in the dermis and in other parts of the limb are the main source for the blastema formation.

4.2 Dedifferentiation observed

For dedifferentiation to take place, certain observable histological events should occur. For example, dedifferentiation of muscle should involve separation of muscle syncytia into mononucleated cells. Observation is feasible with the use of electron microscopy, and the fate of these cells could also be traced with the use of radioactive material. Hay and Fischman (1961) provided the strongest evidence that dedifferentiation takes place; they observed the transition of muscle cells to mononucleated cells. These cells oriented themselves, moved toward the blastema, and showed ribonuclear granules and large nucleoli, indicating active protein synthesis (see also Hay, 1966). These events had also been seen by Bodemer and Everett (1959) and by Anton (1965) with the use of ^{35}S-methionine. This argues against the idea of proteolysis of these cells, because cells under proteolysis do not need active protein synthesis. The cells were also characterized by the presence of many vesicles and the movement of the mitochondria to one end of the cell. Injection of tritiated thymidine in combination with electron microscopy enabled the authors to identify and trace the cells that were labeled (Table 4.1). The data obtained from these experiments conclusively demonstrated that the blastema cells are produced by dedifferentiation.

These studies disputed the epidermal origin. When the limbs were injected with tritiated thymidine two days prior to amputation, only the epidermis was labeled and the label was found later in the wound epithelium but not in the underlying mesenchyme (see also Hay, 1966). Earlier studies by Brunst (1950), Rose, Quastler, and Rose (1955), and Rose (1962) have supported the idea that the epidermis participates in normal regeneration. When irradiated limbs were provided with normal epidermis, regeneration of the irradiated limbs was restored. It is quite possible that viable subepidermal cells had accompanied the epithelium and could, therefore, account for the regenerates (Chalkley, 1956).

Dedifferentiation has also been histologically studied with the use of two enzymes, alkaline phosphatase (Karczmar and Berg, 1951) and

Table 4.1 Origin of blastema cells in the newt *Triturus viridescens*. An autoradiographic study by Hay and Fischman (1961)

INJECTION OF ^{3}H-THYMIDINE

	Before Amputation	During Regeneration	During Regeneration
		5, 10, 15 days fixation a day after	1-28 days fixation same day
RESULT:	Only epidermis, blood and wound epithelium were labeled.	Dedifferentiated tissues and subsequent blastema cells were labeled.	DNA synthesis start 4-5 days post amputation in dedifferentiated muscle and tissue near the amputation site. Highest increase 10-20 days. Wound epithelium high at 8 days. Later only 2% of the cells are labeled.
CONCLUSION:	Epidermis does not participate in blastema formation.	No histolysis, but dedifferentiation happens. Blastema cells are produced by dedifferentiated tissue.	Blastema cells are produced by the inner dedifferentiated tissues.

acid phosphatase (Schmidt and Weary, 1963; Ju and Kim, 1994). Activity of both of these enzymes is high during the dedifferentiation stage. Alkaline phosphatase appears 3 days after amputation in the hypodermal fibroblasts, the dedifferentiating muscle, and cartilage cells, and remains strong during the formation of the blastema of the axolotl. Acid phosphatase is present in the apical wound epithelium, the dedifferentiating tissues, and the blastema cells of the newt *D. viridescens* and the Korean newt *Hynobius leechii*. Such assays have also helped to study the degree of dedifferentiation from different amputation levels (Ju and Kim, 1994). When a limb is amputated at a distal level (zeugopodium) or proximal level (stylopodium) the extent of dedifferentiation is not the same. In fact, the more proximal the amputation is performed, the more extended the degree of dedifferentiation. At 10 days after amputation, acid phosphatase activity is 30 percent higher in the proximal blastema than in the distal one. This indicates a level of dedifferentiation that is 30 percent more extended in the more proximal regions.

4.3 Transplantation experiments

Cell dedifferentiation and differentiation during blastema formation were also studied by transplantation experiments. Early experiments by Patrick and Briggs (1964), where triploid cartilage cells were transplanted into diploid hosts, gave no evidence of metaplasia of cartilage during regeneration. Namenwirth (1974) and Dunis and Namenwirth (1977) transplanted triploid cells into the limbs of diploid X-irradiated host animals. Since irradiation inhibits regeneration, only the transplanted cells will be able to participate. It was observed that cartilage cells dedifferentiate and subsequently give rise to cartilage, perichondrium, joint connective tissue, subepidermal fibroblasts, and fibroblasts in skeletal muscle. Following dedifferentiation and mitosis, whole muscle was able to give rise to all mesodermal tissues of the limb, including cartilage. Epidermis was found to produce only epidermis. Steen (1968, 1973) used similar techniques to show the contribution of muscle to cartilage and cartilage to connective tissue (Figure 4.1). Dermis also was found able to contribute to the blastema cells. Triploid skin was transplanted to denuded limbs of diploid axolotls that had received X-irradiation. Cartilage, perichondrium, joint and general connective tis-

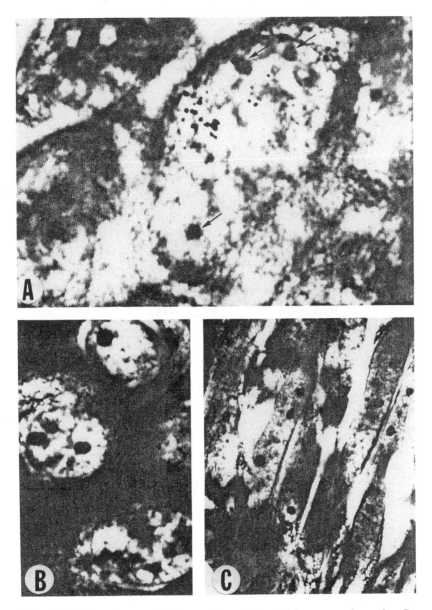

Figure 4.1 Chondrocyte differentiation originated from muscle grafts. In this experiment, triploid muscle was labeled with ³H-thymidine and was grafted to diploid axolotl. Differentiation of muscle was followed in the regenerates. A: Differentiated chondrocytes with the three nucleoli (arrows) as well as with silver grains found in the wrist of the regenerate. B: Labeled chondrocytes with three nucleoli found in the humerus. C: A differentiated muscle cell with three nucleoli originated from the same graft that gave rise to cartilage. (From Steen, 1973.)

sue, dermis and epidermis were present in the regenerated limbs, but less than 10 percent of them contained muscle that originated from the triploid grafted skin. Similar studies, however, have produced evidence that muscle and possibly melanocytes can be differentiated from cartilage (Maden and Wallace, 1975; Desselle and Gontcharoff, 1978). It seems, therefore, that non-cartilage connective tissue or myogenic cells can produce cartilage, but there is no evidence yet that muscle cells can be transformed by non-cartilage connective tissue.

The ability of a given tissue (that is, muscle) to undergo dedifferentiation and give rise not only to muscle but also to other tissues (such as cartilage) implies metaplasia. Such a case had been shown early on by Bischler (1926) and Thornton (1938a) (see also Vorontsova and Liosner, 1960). Thornton amputated limbs at the elbow level and subsequently removed the humerus. Limb regeneration took place with all the skeletal elements distal to the amputation present (Figure 4.2). Trampusch (1956) and Trampusch and Harrebommee (1965) have also shown possible metaplasia from bone. In his experiments, he transplanted the humerus from a forelimb to an amputated hindlimb after the hindlimb was completely irradiated and the femur removed. Regeneration proceeded, indicating that the transplanted bone restored the regenerative capacity of an irradiated limb, which was otherwise unable to regenerate. These experiments indicated that cells in the regenerate can be derived by metaplasia (Figure 4.3).

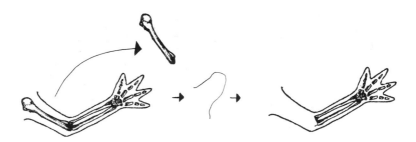

Figure 4.2 A classical experiment indicating metaplasia in the regeneration of the skeleton. The humerus is removed from an intact limb and the limb is amputated through the area of the existing humerus. The regenerating limb contains all the skeletal elements distal to the humerus even though the humerus was not present to be the source of them. (Thornton, 1938a; illustration adapted from Casimir et al., 1988b.)

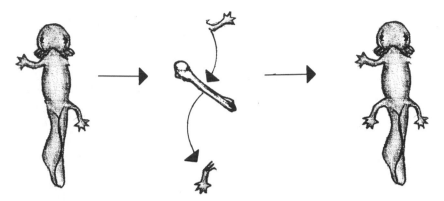

Figure 4.3 An illustration representing the effects of bone transplantation in irradiated host limbs. The humerus was removed from the irradiated limb. In exchange, humerus was transplanted from a nonirradiated donor. The irradiated limbs receiving the transplant were able to regenerate after amputation was performed through the transplant. This indicates that the normal transplants can rescue the irradiated limbs, which otherwise would not regenerate. Furthermore, these experiments indicate that the bone (or similarly skin) can give rise to every other tissue in the limb possibly by metaplasia. (After Trampusch, 1956.)

4.4 The cellular and molecular evidence

Metaplasia has also been shown by using cells grown in culture. Usually, limb cells can grow in culture (see Chapter 8) when an explant of blastema or muscle is placed on a petri dish. The mononucleated cells that come out of such explants express myf-5 (Tsonis, Washabaugh, and Del Rio-Tsonis 1995), can grow well, and a few days later can start forming polynucleated myotubes. Such myotubes were microinjected with specific dyes and treated with ^{3}H-thymidine, and their fate was followed after transplantation into the limb blastema. It was found that such cells can dedifferentiate because the dye was later found in mononucleated cells. In fact, these cells had their nuclei filled with silver grains, indicating active DNA synthesis. Moreover, it was also shown that cartilage cells in the regenerate were labeled. This provides evidence of dedifferentiation and subsequent metaplasia during limb regeneration (Figure 4.4) (Lo, Allen, and Brockes, 1993). Results from mammalian systems provide some insights into the role of cyclins in dedifferentiation. When terminally differentiated myotubes were

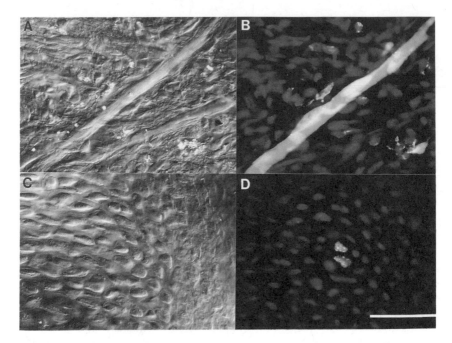

Figure 4.4 Muscle fiber and cartilage differentiation resulted from transplantation of myotubes. In this experiment, cells grown out of muscle explants were allowed to differentiate to myotubes. These myotubes were then labeled with rhodamine-conjugated lysinated dextran by microinjection. The myotubes were then implanted into amputated newt limbs and their fate traced. A: Labeled muscle fiber 9 days after implantation seen under differential interference optics. B: Rhodamine labeling (of A) in the cytoplasm, indicating that the muscle fiber was formed from the implanted myotubes. C: Differentiating cartilage 26 days after implantation. D: Same as in C under fluorescence showing two labeled cartilage cells. These cells have arisen from the implanted myotubes by transdifferentiation. Bar is 0.1 mm. (Courtesy J. P. Brockes.)

immortalized by the large T antigen, they initiated DNA synthesis. DNA synthesis was correlated with cyclin A and its receptor, the kinase cdk2 (Cardoso, Leonhardt, and Nadal-Ginard, 1993). These data might indicate that key regulatory molecules such as cyclins (important for the cell cycle) might be activated by protein kinases. In fact, this scenario would fit with the idea of second messenger activation reported in Chapter 2. Similarly, the tumor suppressor retinoblastoma protein (Rb) is required for the terminal differentiation of mammalian muscle (Schneider et al., 1994) and might be involved in such regulation in the amphibian limb.

The development of monoclonal antibody production technology

Figure 4.5 Staining of a newt blastema 8 days post-amputation with the mAb 12/18, showing staining in the dedifferentiating tissues and the accumulating blastema cells. Epidermis is colored in white. Bar is 0.5 mm. (Courtesy J. P. Brockes.)

enabled scientists to produce good markers for cell lineages. Likewise, Kintner and Brockes (1984) applied this technology and received a number of interesting monoclonal antibodies directed against the regenerating tissue of the newt. One of them, the so-called 22/18, identifies blastema cells (Figure 4.5) as well as myogenic cells (Griffin, Fekete, and Carlson, 1987), and another, 12/101, recognizes only myofibers (Figure 4.6). However, a small population in the regenerate has both the blastemal and the myofiber antigens. Such a finding supports the idea that blastema cells originate, at least in part, from dedifferentiated myofibers. The 12/18 antibody also recognizes Schwann cells. This might explain the "paradoxical" regeneration where it is believed that the cells in the regenerate can be derived from these cells (see Section 4.5). Studies on the identification of the interacting antigen showed that the antibody 12/18 recognizes a conformational change in an intermediate filament during limb regeneration (Ferretti and Brockes, 1990).

Casimir et al. (1988a) studied this phenomenon using molecular

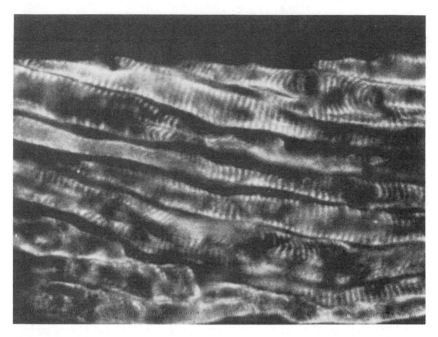

Figure 4.6 Staining of normal myofibers with mAb 12/101 showing the striated patterns. (From Griffin et al., 1987.)

probes. The probe was the newt cardioskeletal myosin heavy chain gene, which shows two tissue-specific hypomethylations. One of them is specific to muscle lineage (Hypo A) and the other to the connective tissue component of muscle (Hypo B). DNA was isolated from three different tissues in the regenerate: regenerate cartilage, regenerate cartilage from limbs with their humerus removed, and blastema. The DNA was probed with the myosin gene and the bands were compared with the ones from normal cartilage. Both markers showed an increase in the regenerating tissues, suggesting that the cells that carry the marker contributed to the blastema and cartilage of the regenerate. Especially pronounced was the contribution of the Hypo B (Figure 4.7). These results show that cells of the muscle lineage can dedifferentiate and contribute to the blastema. Moreover, these results gave the first molecular evidence of metaplasia because, when the humerus was removed, the regenerated cartilage cells carried the Hypo B marker, indicating selective recruitment from the "connective tissue" into cartilage (Casimir et al., 1988b).

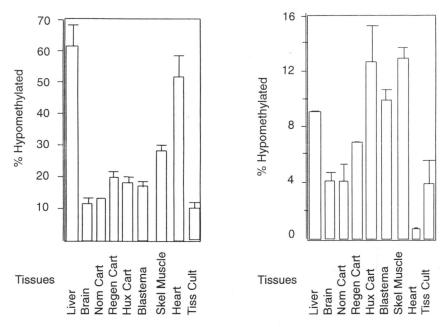

Figure 4.7 Behavior of marker hypomethylations in regeneration. These histograms show the level of hypomethylation for two markers (Hypo A, muscle marker, left panel; Hypo B, connective tissue marker, right panel). The results were derived from DNA hybridizations; hypomethylation was quantified in a number of different regenerate and normal newt tissue DNAs. Hux cart: cartilage from regenerated limb from which the humerus was removed at the time of amputation. Blastema: cone stage blastema. Tiss cult: cultured newt muscle cells. Note the increased representation of markers in regenerative tissues relative to normal cartilage. The increase is especially dramatic for Hux cart with Hypo B. (Adapted from Casimir et al., 1988b.)

4.5 Paradoxical regeneration

Usually, irradiated limbs do not regenerate. Denervated limbs (see Chapter 6) are also unable to do so. Under certain circumstances, however, if limbs are denervated shortly after irradiation and before amputation, they do regenerate. Such a regeneration was termed "paradoxical" by Wallace (1972). Maden (1977a) extended the studies on paradoxical regeneration. In experiments where only the hand was shielded, regeneration took place, whereas if the whole limb was irradiated no regeneration was observed. This was interpreted by proximal migration of cells from the hand in response to denervation. These cells were able to form a blastema. Histological observations showed that when the limbs were totally irradiated and denervated (so that paradoxical regeneration is not permitted), regrown axons remained unmyelinated. When paradoxical regeneration was permitted by shielding of the hands, the axons became myelinated. Taking into account the properties of Schwann cells for long-range migration, it was concluded that these cells were the only source of blastema that resulted in paradoxical regeneration (Maden, 1977a). These results imply that cells of ectodermal origin (Schwann cells) can undergo metaplasia and form the mesodermal tissues of the limb.

4.6 Cells that contribute the most

From Hay's experiments (Table 4.1) it was clear that the epidermis does not contribute to the formation of the blastema cells. The same is true for the blood cells. These experiments seem to indicate that many different cell types are capable of contributing cells to the blastema under experimental conditions. It seems likely that different cell types can be recruited as needed and that blastema formation can be achieved in different ways. But which cells do contribute the most? The experiments by Dunis and Namenwirth (1977) suggest that fibroblasts from the dermis contribute predominantly to the blastema. Such a conclusion was also reached by Muneoka, Fox, and Bryant (1986). Maden (1977a), nevertheless, has proposed that under certain conditions, the blastema in the axolotl can be completely derived from Schwann cells (see Section 4.5).

4.7 The role of the skin

As discussed, the epidermis does not seem to contribute to the formation of the blastema. However, the covering of the wound epithelium is important for regeneration to be permitted. If skin is removed from a limb, the lost tissue is covered by the epidermis. Healing can be prevented if the skinless portion is grafted to an internal body site, such as the flank musculature or the coelomic cavity (Goss, 1956a). If such a limb is amputated, the stump is covered by the peritoneum but regeneration does not proceed. Therefore, the influence of the epidermis is permissive, but the epidermis bears no instructions as to what will be regenerated. Related to this, it should be mentioned that if skin is used to cover the amputation surface, regeneration fails to proceed (see Rose, 1970; Mescher, 1976). The underlying dermis may behave as a barrier that prevents the appropriate epithelial–mesenchymal interactions from taking place.

Studies stemming from the so-called limb territory provide interesting ideas about the role of the skin in regeneration. By "limb territory," we mean the area where limb regeneration can occur. The ability of nerve deviation in an intact limb to induce supernumerary limbs has been employed to map the limb territory. When a nerve from the arm is deviated to a wound in the shoulder, a supernumerary limb can be induced. If the nerve is deviated far away from the limb territory, no regeneration occurs. But if full-thickness skin from the limb territory is grafted to that area, a supernumerary limb can be achieved (Figure 4.8).

The role of skin in regeneration of irradiated limbs has demonstrated that skin, as well as bone, will restore regeneration of these limbs (Trampusch, 1956). When skin from a normal limb or even skin from the head visceral or embryonic tissues is transplanted in an irradiated limb, regeneration was restored after amputation of the irradiated limb through the graft (Kiortsis, 1953, 1955; Rahmani and Kiortsis, 1961; Lazard, 1967; Goss, 1969).

4.8 Stem cell versus dedifferentiation

The evidence for dedifferentiation, as described in Section 4.2, is indeed overwhelming. However, there is also the idea that reserve cells are the contributors of the blastema in the regenerate. It has been established that muscle regeneration in mammals and in newts occurs through satellite cells (Cherkasova, 1982; Cameron, Hilgers, and Hinterberger, 1986). However, the role of these cells or any other reserve cells in producing the different cell types in the regenerate has not been substantiated. If reserve cells do contribute to the regenerative blastema, then we should circumvent arguments that support the dedifferentiation idea, for example the local origin of blastema cells. An experiment designed for this purpose involved transplantation of triploid muscle into the shoulders of diploid animals and amputation at the elbow level. The

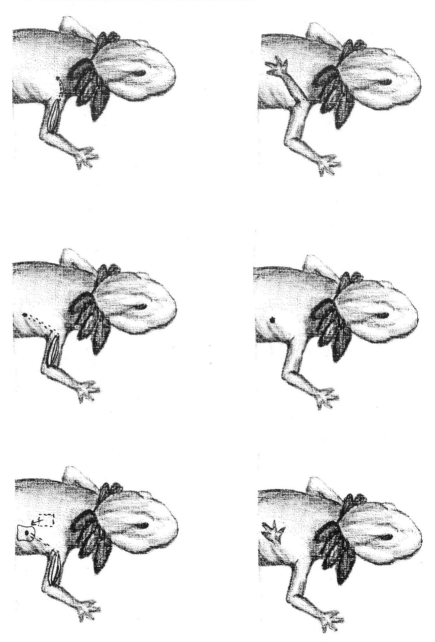

Figure 4.8 An illustration representing the experiment that resulted in the map of limb territory. A nerve from the limb is deviated. This can result in limb formation only within a certain area (upper planes); this is the limb territory. Outside the limb territory a nerve deviation can induce limb formation only if skin from the limb territory has been implanted at the deviation point (lower planes). (Adapted from Goss, 1969.)

experiment demonstrated that triploid cells were found in the regenerating forearm and hand. This indicated migratory capabilities of the implanted cells and argues against the local origin of cells that participate in regeneration (Hinterberger and Cameron, 1991). While this experiment gives support to the contribution of non-local cells, the door to dedifferentiation and metaplasia is still open. Hinterberger and Cameron argue that cartilage dedifferentiation and subsequent differentiation to muscle could be the result of muscle recruitment due to the injury imposed to the limb by the transplantation procedures. Alternatively, the critics of these experiments claim that, at the early stages of regeneration, myoblasts cannot be distinguished from fibroblasts, and therefore metaplasia is ambiguous. However, when cloned chicken myoblasts are grown on a source of chondrogenesis-promoting factor they can form cartilage, and that is strong evidence of such cells' potential for metaplasia (Nathason et al., 1978).

Despite the success in isolating the stem cells for the muscle or bone cell lineages that have convincing roles in the repair of these tissues (Caplan, 1991), the search for reserve or stem cells capable of forming a blastema has not taken off the ground yet. On the contrary, dedifferentiation has been shown convincingly by many different experiments and manipulations, and has received much support through the recent use of cultured myotubes. How can we reconcile all these data that show active roles and participation of many different cell types with the fact that regeneration is feasible when depletion of one or more types occurs? The regeneration of the urodele limb is a very dynamic process, and under normal circumstances it could take place without much regulation. In this simple model, muscle provides muscle, bone provides cartilage, nerves provide nerves, and so on. But when one or more tissues are forced to abandon their role, either by removal or by irradiation, then a rescuing tissue assumes a more loaded role. In other words, depending on how the system is forced, it could use alternative ways and could recruit a particular cell type more than would normally be expected. The plasticity of the system must be remarkable. We do not understand how such a function can take place, that is, how a tissue knows that bone is not present and that it now needs to make cartilage as well. It probably involves signals and communication between different tissues. The presence of stem cells could solve all this, but it seems that even if stem cells exist in the newt limb, they are simply part of the same network.

4.9 The extracellular matrix in the blastema

Terminally differentiated tissues have their own histological characteristics and integrity. Their stabilization is secured by the interweaving
of all cells to the extracellular matrix. Proteins of the matrix will form
polymers or interact with other proteins or receptors on the cell surface.
These specialized cell-to-cell interactions give to the cells of the same
tissue a certain idiosyncrasy and render the tissue rigid and functional.
The extracellular matrix varies from tissue to tissue depending on these
functions. Cartilage, for example, contains much more extracellular
matrix and, as a result, the intercellular space is much larger than, say, in
muscle.

For breakdown of the tissues and dedifferentiation to take place, it
is conceivable that major alterations in the extracellular matrix should
occur as well. Cells need to move, and this can be possible only by
down-regulation of some proteins or up-regulation of other proteins
involved in the extracellular matrix and cell-to-cell interactions. As a
result, we should expect to find interesting regulation in the extracellular matrix of an amputated limb.

The role of proteinases, especially metaloproteinases, has not been
studied in detail during the process of dedifferentiation and blastema
formation. It is assumed, however, that they should play an active role in
tissue remodelling. The regulation of some proteins of the extracellular
matrix and of receptors for the components of the extracellular matrix
has been studied. We now know that some of these molecules are specifically up-regulated and some down-regulated during the remodelling
that leads to the dedifferentiation.

4.9.1 Matrix metaloproteinases

The role of one such proteinase (MMP 9; gelatinase B) has been studied
during axolotl limb regeneration. MMP 9 is up-regulated as soon as 3.5
hours after amputation, becomes high with dedifferentiation, and drops
with differentiation. The strongest activity can be seen in the stump.
Weak activity is seen in the regions distal and proximal to the stump.
MMP 9 is also up-regulated in flank wounds by 1 day after injury, but its
activity falls thereafter. This indicates a more direct role of MMP 9 in
dedifferentiation and epimorphic regeneration (Bryant, 1994).

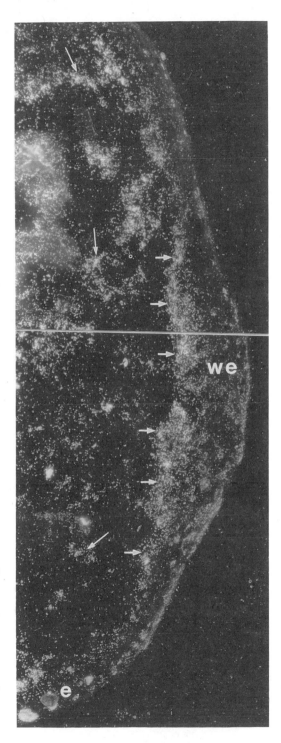

Figure 4.9 Dark-field micrograph illustrating the distribution of type XII collagen transcripts at the pre-blastema stage of newt forelimb regeneration using *in situ* hybridization. Note that type XII collagen mRNA is detectable in the mesenchyme cells at the distal tip of the limb stump (long arrows) and in the basal cells of the wound epithelium (short arrows), indicating a dual source of this protein during regeneration. (Courtesy R. A. Tassava.)

4.9.2 Collagen

Collagen protein has been found to be enriched in the regenerates during the onset of differentiation and chondrogenesis and during the initial period of elongation (Mailman and Dresden, 1976). Collagenase, on the other hand, appears during the stage of dedifferentiation and is localized in the distal region of the stump (Grillo et al., 1968). Several collagenolytic enzymes are up-regulated within hours post-amputation (Dresden and Gross, 1970). Production of one of them, the 92×10^3 M_r collagenase (type IV), was enhanced dramatically during dedifferentiation and blastema formation but was returned to basal levels by the palette stage when differentiation took place (Yang, Huynh, and Bryant, 1993). It seems that degradation of collagen is needed to ensure dedifferentiation. With the use of a monoclonal antibody (MT2), a cDNA clone has been isolated coding for collagen type XII. *In situ* hybridizations show expression in the mesenchymal blastema and in the basal layers of the wound epithelium (Figure 4.9). Expression of collagen type XII is restricted to regenerating limbs, since the developing limb does not express this molecule even after amputation. In addition, type XII expression is nerve independent since its expression does not cease after denervation (Tassava, 1993; Wei and Tassava, 1994). Collagen degradation has also been observed in tissue culture of blastema cells. When blastema cells from the axolotl are placed in high-cell-density cultures in dishes coated with collagen, extensive degradation takes place within 3 days. This property of blastema cells is quite unique and might be related to the matrix remodelling and dedifferentiation seen during amphibian limb regeneration (Groell, Gardiner, and Bryant, 1993).

4.9.3 Laminin

Laminin is a major component of basement membranes. In the intact limb, laminin can be seen in the membranes of the muscle, nerves, blood vessels, skin epithelium, and the glands. Upon amputation, laminin ceases to be detectable in the dedifferentiated tissues (Gulati, Zalewski, and Reddi, 1983). The wound epithelium, especially the cells in close proximity to where the presumptive basement membrane will be built, produces laminin as well as collagen type IV (Del Rio-Tsonis, Washabaugh, and Tsonis, 1992). Laminin was also observed around the

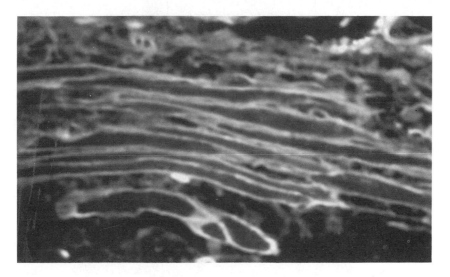

Figure 4.10 Laminin expression around the regenerated myofibers. This expression characterizes the synthesis of the muscle membranes.

regenerated myotubes (Figure 4.10). As in the case of collagen, laminin down-regulation seems to correlate with the onset of dedifferentiation.

4.9.4 Fibronectin

Unlike laminin and collagen, fibronectin is very abundant in the regenerating blastema, and it persists during blastema growth, aggregation, and the early stages of differentiation. As redifferentiation proceeds, fibronectin disappears from the cartilage matrix and the myoblast fusion zone (Gulati et al., 1983). This distribution has been studied with both rat and newt antibodies. A cDNA probe for newt fibronectin has been isolated with the use of a monoclonal antibody (MT4). Expression studies demonstrated that fibronectin is expressed by the regenerating blastema as well as by the developing limb-bud cells (Figure 4.11) (Tassava, 1993).

4.9.5 Tenascin

Tenascin also is up-regulated owing to the regeneration process. It is first seen 2 days after amputation in the wound epithelium and then in the distal mesenchyme 5 days after amputation. In the blastema, virtually all

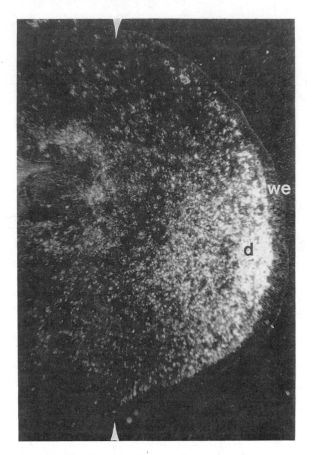

Figure 4.11 Distribution of fibronectin transcripts in a mid-bud stage regenerating newt limb using *in situ* hybridization. Note that fibronectin mRNA is
most highly concentrated at the distal tip (d) of the mesenchyme. The basal
cells of the wound epithelium are also expressing the fibronectin gene. Thus
there are two sources of fibronectin in the regenerating limb: the blastema cells
and the basal cells of the wound epithelium. The arrowheads indicate the level
of amputation. we: wound epithelium. (Courtesy R. A. Tassava and J. Nace.)

mesenchymal cells express tenascin; its expression is also associated with
the redifferentiation process and growth (Figure 4.12) (Onda et al., 1991).

4.9.6 Hyaluronate

Synthesis of hyalunorate begins 2 days after amputation and increases
rapidly during dedifferentiation. The synthesis decreases as the blastema

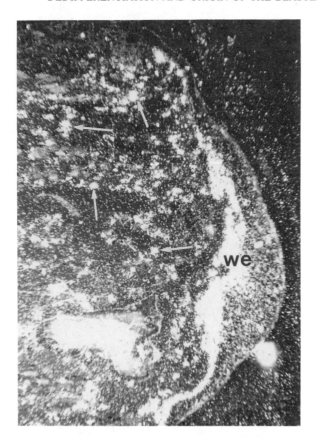

Figure 4.12 Dark-field micrograph indicating the distribution of tenascin transcripts at the pre-blastema stage of newt limb regeneration using *in situ* hybridization. Note that tenascin transcripts are present in the mesenchyme cells of the distal tip of the limb stump (long arrows) and in the basal cells of the wound epithelium (we). (Courtesy R. A. Tassava and H. Onda.)

enters the differentiation phase and growth. This correlates with the activity of hyaluronidase, which appears during differentiation and remains active until 40 days after amputation when maximal chondroitin sulfate synthesis and cartilage differentiation take place (Toole and Gross, 1971; Smith, Toole, and Gross, 1975).

4.9.7 Integrins

The expression of different chains that constitute the integrins shows interesting patterns of expression; alpha3, alpha6, alphav, and beta3 are

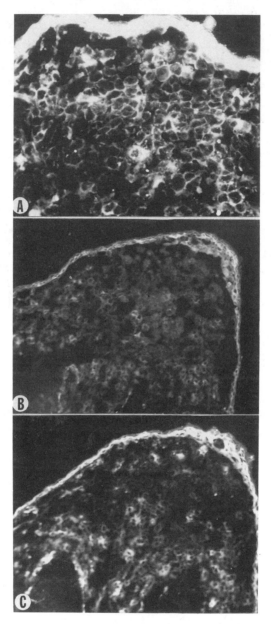

Figure 4.13 Expression of integrins in the regenerating axolotl limb 8 days post-amputation. A: alpha3 expression in the blastema cells. Note high expression in the distal tip of the blastema. B: alpha1 expression in the blastema. Note that expression is down-regulated in the distal tip of the blastema. C: Expression of beta1 in the blastema. Note that expression is down-regulated in the distal tip of the blastema which represents the dedifferentiated blastema cells.

up-regulated during blastema formation in the axolotl. However, alpha1 and beta1 are down-regulated in the regenerate (Figure 4.13). These two are again seen in the palette stage. Interestingly, Alpha1 and beta1 are down-regulated during lens regeneration as well. Alpha1–beta1 integrin is the receptor for laminin. Beta1 integrin has been implicated in the arrest of cell cycle (Meredith Jr. et al., 1995). Down-regulation, therefore, of this integrin might be necessary to allow blastema cells to divide.

4.9.8 Catenins

Catenins are proteins associated with N-cadherins, Ca^{2+}-regulated glycoproteins on the cell surface that play important roles in cell-to-cell interaction and cell adhesion (Wheelock and Knudsen, 1991). The use of an antibody that is reactive to newt tissues on Western blots has provided information on regulation of these associated proteins. Specifically, appearance of one of them can be detected in the regenerate 2 weeks post-amputation (Figure 4.14).

4.9.9 N-CAM

Infusion of N-CAM antibody (cell adhesion molecule) to the regenerating newt limb resulted in growth inhibition. Obviously this molecule is important for blastema organization and growth (Maier et al., 1986).

4.9.10 Other molecules involved in the extracellular matrix

An antibody has been recently described that is directed against a transmembrane molecule, which is involved in cell adhesion. This molecule disappears during the formation of the limb blastema as well as during regeneration of the lens from the dorsal iris (Imokawa et al., 1992; Imokawa and Eguchi, 1992). Disappearance of this antigen seems to be not only necessary but also sufficient for regeneration. Forced masking of the antigen by treatment with the antibody provides the ventral iris with regenerative abilities that are not available otherwise (Eguchi, 1988). A different antibody with similar patterns of disappearance during blastema formation has been reported by Yang, Shima, and Tassava (1992). The ST1 antigen is abundantly present in the normal differenti-

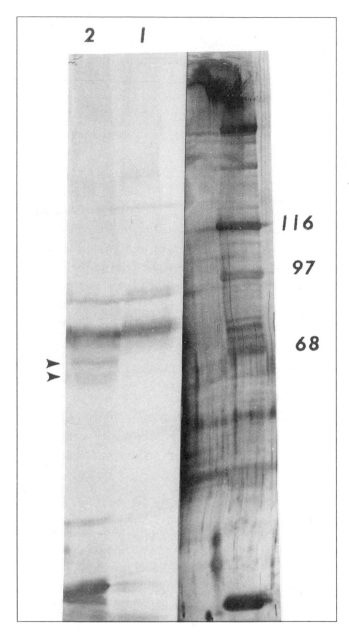

Figure 4.14 Expression of beta-catenin during blastema formation. Lane 1 proteins isolated from intact limb. Lane 2 protcins isolated from 2-week-old blastema. The proteins were run under reduced conditions. Note the appearance of additional bands (arrowheads), cross-reacting with the antibody, in the blastema only. The lane on the right is markers; the numbers represent molecular weight in KDa.

ated tissues, but it gradually disappears during dedifferentiation and remains absent until differentiation (Figure 4.15).

It is obvious from the studies on the ECM that extensive remodelling occurs during the dedifferentiation phase and blastema formation. The coordinate down-regulation of some proteins and the up-regulation of others provide a well-orchestrated pattern by which dedifferentiation

Figure 4.15 Reactivity of monoclonal antibody ST1 to the stump tissues of an adult newt forelimb. The blastema is not positive. Note that breakdown of the ST1 antigen earlier in regeneration formed a concavity into the stump (arrow heads). The level of amputation is indicated by the most distal glands (arrows). The distal tip of the wound epithelium (we) is the distal end of the blastema. (Courtesy R. A. Tassava.)

can take place. Determining the process of this regulation could be all we need to learn about regeneration. It would be of great interest to the field to know and follow this regulation in other animals and correlate it with different regenerative abilities.

4.10 Metabolic changes during blastema formation

Several metabolic and physiological parameters have been studied during the early stages of limb regeneration. This work has been summarized by Needham in his book *Regeneration and Wound Healing* (1952), and events are shown in Table 4.2. It is obvious that the process of limb regeneration results in certain physiological changes in the animal, such as decrease of the pH and of the oxidation–reduction potential. Oxygen consumption decreases at the early stages of blastema formation and increases at the later stages of repair. Similarly, other changes in enzyme content are pronounced. For example, proteolytic enzymes (cathepsins) are at a maximum level 3 days after injury (see also Schmidt, 1966, 1968). This may very well coincide with the extracellular matrix remodeling. Acid phosphatase activity is at its maximum during blastema formation, and this also could be an indicator of dedifferentiation (Ju and Kim, 1994). Other enzymes that are involved in metabolic pathways, such as lactic acid, increase during blastema formation as well.

4.11 Dedifferentiation, regeneration, and cancer

The unique phenomenon of dedifferentiation has its parallel only in the cellular transformation of a healthy cell to a cancerous one. The blastema resembles neoplasms in many ways, as outlined by Prehn (1971): (1) They both arise as a result of trauma; (2) there is cellular proliferation and penetration of nerve fibers into the epithelium; (3) both depend on a nerve supply; (4) both exhibit rapid growth; (5) both show lack of differentiation (temporary in the blastema); (6) both show ability to differentiate to many tissues. The obvious similarity, dedifferentiation, has led many scientists in the past to speculate about cancer induction in amphibians. It is true that spontaneous neoplasia is not a common observation in amphibians. Interestingly, there have been more

Table 4.2 Timetable of metabolic events in regeneration and wound healing in amphibia. (All times in days.)

A. REGRESSION-PHASE

Event and animal	Time of onset	Value at onset	Time of	Value at	Time of	Value at
	(Normal Value)		Maximum or minimum		End of the phase	
Cathepsin activity	---	---	3	maximum	14	---
pH changes:	---	7.2	6	6.6	18	7.2
Eh oxidation-reduction potential	---	225 mv.	10	180 mv.	30	225 mv.
Oxygen consumption	---	Q O_2 = 1.38	8	Q O_2 = 1.02	see Table 3B	see Table 3B
Respiratory quotient	---	1.02	8-20	0.59	see Table 3B	see Table 3B
Concentration of SH groups:	---	20 mg. %	5-10	45 mg. %	60	41.5 mg. %
		40 mg. %	10	55 mg. %		
Acid phosphatase activity:	---	---	15-20	maximum	---	---
Lactic acid concentration:	---	9 mg. %	16	16 mg. %	---	---
		18 mg. %	---	39 mg. %	---	---

B. REPAIR PHASE

Event and animal / Changes in	Time of	Value at	Time of	Value at	Time of	Value at
	Onset		Maximum or minimum		Completion	
PNA concentration	---	---	11-15	maximum	---	---
Free amino acids, concentration changes	---	16.8 mg. %	16-20	35.1 mg. %	---	---
Dipeptidase activity	---	---	10	maximum	---	---
Lability of proteins	---	---	38	maximum	---	---
Oxidized glutathione, concentration changes	24-6	---	---	maximum	60	---
Vitamin C, concentration changes:	---	---	10-60	maximum	---	---
Alkaline phosphatase activity:	5	---	10 (spring) max. 20-5 (summer) max.	10 (spring) max. 20-5 (summer) max.	---	---
Oxygen consumption	8	Q O_2 = 1.02	26-29	Q O_2 = 2.15	---	Q O_2 = 1.38

cases reported for anura that have lost their regenerative ability, than for urodeles (Tsonis and Del Rio-Tsonis, 1988). In addition, numerous studies in the past have demonstrated the difficulty in inducing tumors in salamanders using chemical carcinogens, especially in the regenerating tissues (Zilakos et al., 1992). Rather, administration of carcinogens in the regenerative tissues has sometimes resulted in normal morphogenesis such as generation of accessory limbs (Tsonis and Eguchi, 1981; Tsonis, 1983) or a lens from the ventral iris, a place where lens regeneration never occurs normally (Eguchi and Watanabe, 1973). Tumor formation has been achieved in the limbs of frogs, but only following denervation (which lessens the regenerative ability of the animal) (Outzen, Custer, and Prehn, 1976). Is it possible that the regenerative fields protect the cells from transformation? Or is it that dedifferentiation induced by the carcinogens made the cell express its inherent ability for regeneration rather than cancer? Do these similarities render the blastema cells "immortal"? This is, of course, a fascinating possibility, and it could provide a basic model for cancer biology and regulation.

5

Differentiation
of the Blastema

Once the blastema has been produced there is a period of proliferation
without differentiation. The blastema has to achieve a critical mass
before differentiation and restoration of the missing part. The period of
proliferation lasts for about 2 to 3 weeks post-amputation in the adult
newt. This varies for different species, depending on the size and the age
of the animal. Proliferation is then followed by differentiation. Cartilage
is the first tissue to be seen. The different stages of differentiation in
the regenerating limb are outlined in Table 5.1. Histological sections of
representative samples of the regeneration's progress are shown in
Figure 5.1. As outlined by Iten and Bryant (1973), the rate of elongation
is not influenced by the plane of amputation during the dedifferentiation. However, during differentiation and morphogenesis the rate of
elongation is greater from a proximal amputation than from a distal one.
The different stages of regeneration are the same from a proximal or distal amputation; this is valid for digit regeneration as well.

Table 5.1 Length of limb regeneration process in different urodeles.

Organism	Cone Stage Blastema (days)	Palette Stage (days)	2-Digit (days)	Complete Regeneration (days)
Amblystoma mexicanum 30–40 mm 21ºC	6–8	10–12	14–16	24
Amblystoma mexicanum 130–160 mm, 21ºC	14–21	23–30	31–34	40+
Notophthalmus viridescens 60–80 mm, 21ºC	14–16	21–24	28–30	40+
Cynops pyrrhogaster 80–100mm, 21ºC	14–16	20–25	30–35	50+
Taricha granulosa 140–170 mm, 21ºC	19–26	28–32	33–40	150+

5.1 Differentiation and regeneration of muscle

Muscle is a tissue known to regenerate in all vertebrates. Such regeneration, however, is called tissue regeneration, as opposed to the epimorphic regeneration seen in salamanders. In tissue regeneration of muscle there is destruction of myofibrils into individual sarcomeric units and invasion of the traumatized muscle by numerous macrophages. During epimorphic regeneration, the destruction of the muscle fibers involves fragmentation of the distal ends of the severed muscle fibers (Hay, 1971). The muscle is invaded by macrophages, but they come in smaller numbers. It has been proposed that muscle fibers attract macrophages by producing complement factors (Engel and Biesecker, 1982). True enough, complement factor C3 has been located in the blastema and the forming muscle of the axolotl (Tsonis and Lambris, unpublished). Upon destruction of the fiber, the regeneration cell contains a nucleus from the multinucleated myocyte and cytoplasm that were formerly part of the syncytial fiber. The basal lamina is recreated in the regenerated fiber. The original lamina seems to play roles in innervation, and in the proliferation of the myoblasts, by providing substrate and in permitting macrophages to pass through selectively (Thornton, 1938b).

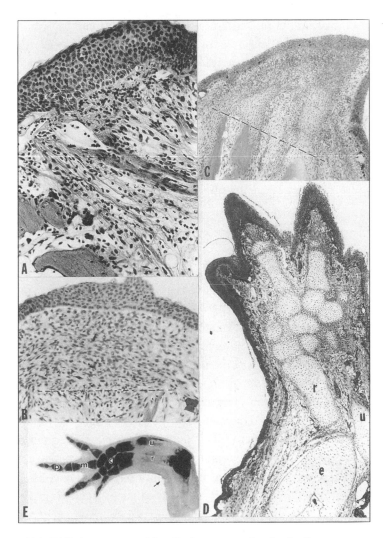

Figure 5.1 Different stages of forelimb regeneration in the Japanese newt *Cynops pyrrhogaster.* A: A section through the stump 7 days post-amputation. Note the wound epithelium and the beginning of dedifferentiation and blastema formation underneath it. B: A section through a blastema 2 weeks post-amputation showing the accumulation of blastema cells. C: A section through a regenerating limb 5 weeks post-amputation. Note the differentiating cartilage composing the ulna and radius. D: A regenerate 2 months post-amputation showing most of the missing part regenerated. Amputation at the level of elbow (e). E: A regenerated limb stained with Victoria blue B. The amputation was performed at the level of the elbow (arrow). u: ulna, r: radius, c: carpals, m: metacarpals, p: phalanges. Dashed lines indicate the level of amputation. The hand is also called autopodium; the ulna and radius region, zeugopodium; and the region containing the humerus, stylopodium.

It seems that the musculature possesses positional information. If individual muscles (that is, flexor and extensor) are cross-transplanted, regeneration of complex multiple structures can be obtained (Carlson, 1974a, 1975a, 1975b). These supernumerary structures can be elicited due to the confrontation of positional values (see Chapter 16). However, if minced muscle is transplanted, muscle regenerates through tissue regeneration first and then normal limb regeneration occurs, except if flexor and extensor muscles are exchanged (Carlson, 1974a, 1975b). Interactions of tissue and epimorphic regeneration can be seen in the axolotl. In an experiment, the muscle of the upper arms was minced, and an amputation was then performed through the minced portion. In so doing, minced muscle (tissue regeneration) was under the influence of dedifferentiation (epimorphic regeneration). It was shown that the minced muscle was able to take part in the dedifferentiation process. In other words, epimorphic regeneration is dominant over tissue regeneration of the muscle (Carlson, 1979).

Myogenic cells have been studied during regeneration in the newt limb with the use of two monoclonal antibodies: the 12/18, which identifies a subset of blastema cells, and 12/101, which is myofiber specific (Griffin et al., 1987). Muscle in one limb was minced while the contralateral was amputated. The 12/101 staining showed that myofiber degeneration took 8 to 10 days and myogenesis started 2 days later. The 12/18 was able to recognize cells in the regenerating minced muscle that became the predominant population. A portion of these positive cells was also stained with 12/101, indicating that the 12/18 positive cells can also be myogenic and can represent new myogenesis occurring by tissue repair (Griffin et al., 1987). The expression of two members of the myogenic regulatory gene family, MRF-4 and myf-5, has been studied during limb regeneration in *N. viridescens* as well. The myogenic gene MRF-4 is expressed in adult muscle, but it is absent in the early blastema. It reappears at the completion of regeneration. These patterns of expression are similar to the ones for muscle-specific myosin gene. In contrast, myf-5 is expressed in muscle but also in the regeneration blastema. While MRF-4 may regulate muscle phenotype and its repression related to dedifferentiation, myf-5 may play a role in maintaining a distinct myogenic lineage during blastema formation (Simon et al., 1994).

Epimorphic regeneration of muscle has been shown to proceed normally when essentially all muscle fibers have been removed from the amputation surface (Carlson, 1972). While this may testify to the potential of the limb to compensate for lost tissues by recruiting others, these results might also be explained by the remaining residual muscle.

The origin of myoblastic cells during the regeneration of muscle has been proposed to be post-satellite cells. Such cells have been observed in *Triturus* by Cherkasova, (1982) and by Cameron et al. (1986) in *N. viridescens*. Even though such cells are responsible for muscle renewal in many vertebrates, including mammals, their contribution to epimorphic limb regeneration has been the subject of debate (Chapter 4).

5.2 Differentiation and regeneration of skeletal elements

During the dedifferentiation stage, osteoclasts and chondroclasts are responsible for removing the distal skeletal material that was produced by the amputation. Next, periosteal cells start depositing cartilaginous matrix around the distal part of the skeleton. This periosteal cartilage forms a cuff around the amputated bone and differentiates to cartilage directly. Once the periosteal cuff has formed a definite and recognizable cartilaginous matrix, the skeletal elements within the blastema start to differentiate. The presence of the cuff by itself does not seem to be a prerequisite for regenerative capacity. The periosteal cuff is made even when regeneration is inhibited and in animals that do not regenerate.

Addition of extra bones does not lead to integration during morphogenesis. Instead, supernumerary skeletal structures form within the regenerate (Goss, 1956b).

6

Nerve Dependence
of Regeneration

Only 55 years after Spallanzani's discovery of the regenerative abilities of the salamander – a short time in the scientific pace of the eighteenth and nineteenth centuries –Todd made a very significant discovery in 1823. He observed that limb regeneration was adversely affected where he interrupted the sciatic nerve. Todd realized that if this interference happens after the wound has healed, regeneration is either inhibited or retarded. No further experiments were reported with the regenerative limb of the salamander for about 100 years. Wolff (1902) was the first to confirm Todd's results and reiterated the importance of nerves in limb regeneration. He found that when limbs of *Triton* were deprived of sensory and motor nerves at the site of amputation, the regenerative capacity was lost (Figure 6.1). It reappeared only upon reamputation months later, after motility and sensitivity had been regained in the limb. This showed that the nerves can be regenerated and effectively participate in limb regeneration. In the following year, Rubin (1903) showed the phenomenon of nerve dependence by resecting the

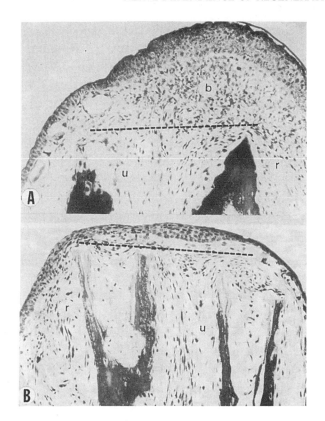

Figure 6.1 Effects of denervation on blastema formation in the newt *Notophthalmus viridescens*. A: A section through a regenerating blastema of an innervated limb 4 weeks post-amputation. Note the normal accumulation of blastema cells. B: A section through a blastema from a denervated limb 4 weeks post-amputation. Accumulation of blastema cells has been inhibited. The dashed line represents the plane of amputation. u: ulna, r: radius.

brachial plexus. Similar results were obtained subsequently by others, including Schotte (1922, 1923, 1926) and Weiss (1922).

Regeneration of the larval limb requires nerves as well. In fact, it seems that the effects of denervation on the amputation stump are more profound in the larval, than in the adult, limb. Unlike the adult, the regeneration stump does not preserve its integrity but instead slowly regresses and finally resorbs completely (Shotte and Butler, 1941; Karczmar, 1946). An intriguing exception to the requirement for nerves was reported by Yntema (1949). He reported that so-called "aneurogenic" limbs, obtained by extirpation of the nervous system in the embryo, can regenerate. This could imply that the nerve dependence of limb regeneration arises as a result of the limb–nerve association during embryonic development. Likewise, absence of nerves may result in abundance of other factors that become crucial for limb development and regeneration. Thornton and Thornton (1970) showed that even these aneurogenic limbs become nerve dependent when transplanted onto normal larvae and innervated by the host nerves. Upon amputation and host nerve denervation, these limbs were able to regenerate up to 11 days after grafting. By the day 13 and thereafter, the limbs were not able to regenerate without nerves, thus behaving like normal innervated limbs. Wallace (1981) partially denervated limbs and showed that the frequency of regeneration was roughly proportional to the number of the remaining nerves. Related to this, Butler and Schotte (1949) showed that in larvae there is a critical, nerve-dependent period, extending to the day 7 to 9 post-amputation, after which regeneration will continue without nerves.

Several experiments have been performed to identify the nerve component that is mostly responsible for the effects on regeneration. The nerve components are motor, sensory, and sympathetic. As early as 1902, Wolff found that the sensory ganglia were the component responsible. However, another school of thought mainly championed by Walter (1919) and Schotte (1923) attributed the dependence to the sympathetic system. Locatelli (1924, 1929) provided further support for the role of sensory neurons. The somatic motor supply was mainly supported by Godlewski (1904). The apparent discrepancies on the role of the different nerve components could have resulted from flaws in operations and lack of staining procedures specific for nerve fibers. In a series of papers published in the 1940s, Singer persistently attacked this problem. Using histoanatomical techniques that allowed him to identify nerve tissue and thus interpret the results of operations (Singer, 1942a, 1942b, 1943, 1945, 1946a, 1946b, 1947a, 1947b), he showed that only the sensory nerves could stimulate regeneration in the absence of the other two. Analysis of the number of nerve fibers (sensory, motor, and sympathetic) contributed by each of the brachial spinal nerves showed that the sensory were most numerous. The ratio was sympathetic:motor:sensory 1:6:22 (Singer, 1946b). This ratio can be correlated with the ability of each nerve component to influence limb regeneration, which means that all components are effective if the quantity is sufficient. Partial destruction of sensory nerves, for example, renders the motor and sympathetic nerves ineffective to a similar degree. Therefore, all the fibers could have the quality necessary for regeneration. In this respect, it should be noted that this could include the fibers of the central nervous system, since fibers of the spinal cord are necessary for tail regeneration.

How does the nerve exert its effect on limb regeneration? It has been hypothesized that the nerve activity on limb regeneration is "trophic," that is, it promotes the accumulation and mobilization of the blastema cells (Singer, 1952, 1978). In this respect, it had been suggested by Guyenot and Schotte (1926) that Schwann cells that migrate along injured axons could invade the amputation site and become the major source of blastema accumulation (see also "paradoxical" regeneration, Chapter 4). Regeneration depends largely on the proliferation of the blastema cells. A cardinal effect of denervation is the arrest of growth of the blastema cells. As mentioned in Section 4.4, a monoclonal antibody was used to provide evidence that Schwann cells and muscle fibers contribute to the blastema (Kintner and Brockes, 1984). Furthermore, this antibody recognizes blastema cells whose division is dependent on the nerve (Brockes and Kintner, 1986). Based on recent information on how nerve growth factors work, it is assumed that those cells are mitogenic. It seems that the nerves supply a crucial factor that promotes proliferation of the blastema cells. Denervation seems to affect the dominant cell population in the blastema, which is the cells that contain extensive endoplasmic reticulum (Bryant, Fyfe, and Singer, 1971). These cells are probably the so-called "pleiomorphic" cells and have been well characterized from *in vitro* work (Chapter 8).

Anura (frogs and toads) lose their regenerative capacity after metamorphosis. Although several theories and ideas have been presented to account for this postmetamorphic loss, not much is known. One idea is that frog nerves provide an inadequate supply of the neurotrophic factor. Experiments conducted by Singer (1951, 1954) support this idea. When the sciatic nerve and its tibial and peroneal branches were dissected and rerouted to the stump of a previously amputated forelimb, regeneration was improved but patterning was still deficient. Additional nerve supply was able to stimulate only minimal regeneration in the lizard limb (Singer, 1961a). Similar experiments have been performed with the chick limb bud. Fowler and Sisken (1982) and Sisken, Fowler, and Barr (1986) implanted neural tubes from stage 15 donors into amputated wing buds of 4-day-old (stage 24) chick embryos. At this stage the limb buds contain only undifferentiated mesenchymal cells. A significant growth response resulted and nearly fully developed limbs were seen. One of the problems here is that the wing buds at stage 24 have not yet formed the skeletal elements and, therefore, it is not clear if this response is regeneration or regulation. The nerves, however, might not be the limiting factor in anurans, because mature stumps of postmetamorphic animals can support regeneration when a blastema from a premetamorphic animal is grafted (Sessions and Bryant, 1988). Studies on the expression of Hox D genes after amputation of the wing bud have indicated that the proximal mesenchyme does initiate the expression of distally expressed Hox genes (Hayamizu et al., 1994) (Chapter 18). Further studies with Hox genes might provide some clues as to the regulative mechanisms underlying responses of limb buds to amputation. At any rate, augmentation of a nerve supply suggests that loss, or downregulation, of a neurotrophic factor could be the reason for decreased regenerative ability in anura and even higher vertebrates. Alternatively, the tissue of limbs from animals unable to regenerate may have become somewhat refractory to the nerve stimulation.

Nerve dependence has also been seen with blastema explants grown *in vitro*. Blastema explants in culture exhibited growth, and differentiation resulted when the explants were innervated but not when they were denervated (Liversage and Globus, 1977).

The nerves can also induce supernumerary limb formation if they are deviated within the limb or the limb field. Locatelli (1929) was able to induce supernumerary limb formation, which she believed always took the form of the limb from which the nerve was derived. Such a belief, however, was challenged by Guyenot and Schotte (1926), who found that the structure of the supernumerary limb was that of the limb nearest to where the nerve was implanted. In fact, Guyenot, Dinichert-Favarger, and Galland (1948) and Kiortsis (1953) used this property of the nerve deviation to study limb "territories" in *Triton*. These investigators were able to discover how far from the limb a nerve could be deviated for a limb to be generated; the area around the limb that supports such limb formation is the limb "territory" (see Figure 4.8). Kiortsis (1953) showed that the only way an additional limb can be formed by nerve deviation outside the limb territory is by transplanting full-thickness skin from the limb territory along with the deviated nerve.

Bodemer (1958) studied this phenomenon and concluded that such supernumerary limbs arise from the dedifferentiated muscle at which the nerve was deviated.

6.1 Mode of action of nerve supply

The inhibition of growth upon denervation suggests that nerves act on the cell-division machinery of the cell. An early experiment by Lebowitz and Singer (1970) showed that protein synthesis is inhibited in the blastema upon denervation and that stimulation of protein synthesis occurred after infusion of nerve homogenates. Similarly, it was later shown that brain extracts were able to restore protein synthesis in denervated blastemata (Singer, Maier, and McNutt, 1976). The brain extract could be from newt or even from chicken (Choo, Logan, and Rathbone, 1978). Studies on DNA synthesis have also provided information on the phases of the cell cycle on which nerves act. DNA synthesis, as indicated by ^{3}H-thymidine incorporation in denervated stumps, resembled the contralateral controls; however, cell division was observed only in the innervated limbs. The explanation of this may be that the limb stump can initiate the regeneration process despite the presence or absence of nerves, but the cells do not proliferate because of a blockade in the G_2 phase of the cell cycle (Tassava and Mescher, 1975; Tassava and Olsen, 1985). Later experiments implied an action of nerves in both G_1 and G_2 phases of the cell cycle. Maden (1978) observed mitosis after denervation (by ^{3}H-thymidine labeling) and he proposed that the cells divide and become arrested at the G_1. Carlone and Foret (1979) supported the idea that blastema cells are arrested in G_2 in the absence of nerves. They observed that once the cells with stabilized mitotic index (cone stage blastema explants) were given cAMP, mitotic activity was observed after 8 hours. This coincides with the duration of the G_2 phase. Similarly, Globus (1978) found mitosis in cultured blastema. Thus, even though denervated limbs do not regenerate, cells nevertheless dedifferentiate and enter the cell cycle, but the limited degree of cycling precludes blastema formation.The control of cell proliferation seems to be very specific in regenerating blastema. In fact, an additional supply of nerves has no effects on cell cycling. Implantation of two ganglia in denervated newt limbs can restore levels of cell cycle activity and stimulate regeneration (Goldhamer, Tomlinson, and Tassava, 1992). However, ganglia added to innervated blastema do not increase the numbers of actively cycling cells, and the rate of regeneration is not enhanced.

Denervation induces a transient increase in DNA, RNA, and protein synthesis followed by significant reduction, as well as a decrease in the number of ribosomes and of nascent polypeptide chains (Bantle and Tassava, 1974; Dearlove and Stocum, 1974; Mescher and Tassava, 1975; Bast, Singer, and Ilan, 1979). Differences in protein synthesis in the newt brachial plexus nerve ganglia have also been indicated. Limb amputation increased protein synthesis fourfold in the ganglia, whereas denervation resulted in a twofold increase. New protein synthesis was not observed, but the differences were quantitative. Specifically, a group of basic proteins ranging from 15,000 to 31,000 Da in molecular weight was expressed predominantly in the experimental ganglia (Bao et al., 1986). This might indicate specific up-regulation of important neurotrophic factors. Amputation-induced changes in protein synthesis were evident in both normal and denervated limb stumps (Garling and Tassava, 1984). However, analysis of protein synthesis in the brains of newts undergoing limb regeneration suggested quantitative and qualitative differences when compared with synthesis from a brain of an intact animal. In fact, differences were observed when the limb or the tail was amputated, indicating that the differences were likely due to the injury or stress (Tsonis, Washabaugh, and Del Rio-Tsonis, 1992). Identification of such proteins will likely provide useful insights on the neurotrophic effect on limb regeneration. Reduction of protein synthesis rate has been reported in both epidermis and the blastema after denervation (Geraudie and Singer, 1978). Normal nerve supply could in fact exert a negative control over some proteins during limb regeneration. For example, it has been demonstrated that c-*myc* protooncogene expression was augmented after denervation. Similar accumulation was observed for the proliferating cell nuclear antigen (PCNA), a subunit of the DNA-polymerase delta that is involved in DNA replication with increased rates at the S phase and for enolase. This accumulation is probably a result of control at the translational level, especially for c-*myc* (Geraudie et al., 1990; Lemaitre, Mechali, and Geraudie, 1992).

The nerve has also been found to change the identity of blastema cells. This was shown with the use of the mAb 12/18 on blastema sections from aneurogenic limbs. Although 12/18 stains the blastema of the innervated limbs, it fails to do so in the blastema from aneurogenic limbs. Since even the aneurogenic limb can regenerate, this means that these blastema cells have altered biochemical characteristics due to the deprivation of the nerve (Fekete and Brockes, 1988).

6.2 Control of cell proliferation

Catecholamines are amines or peptide hormones that mediate responses of very short duration by acting as neurotransmitters. Catecholamines have been found to be abundant in the regenerating tissues of the newt. Inhibition of the enzyme tyrosine hydroxylase by alpha-methyl-p-tyrosine impairs the synthesis of catecholamines and retards limb regeneration, whereas reserpine, which stimulates release of catecholamines from the nerves, enhances the rate of limb regeneration (Taban, Cathieni, and Constadinidis, 1976). Catecholamines have been linked to the activation of adenylate cyclase and the production of cAMP. Levels of cAMP show an abrupt increase during the early bud stage of blastema formation, but the amounts decline below the normal levels of an unamputated limb by the late cone or palette stage (Jabaily, Rall, and Singer, 1975). At the same time, the cAMP phosphodiesterase (the enzyme that degrades cAMP) is low at stages where cAMP is high (Carroll and Sicard, 1980). Initial reports claimed that variations of the endogenous cAMP levels correlated with noradrenaline (Taban, Cathieni, and Schordeter, 1978). Subsequent experimentation, however, failed to confirm this correlation. It was found that even though noradrenaline stimulated cAMP, through beta-adrenergic receptors, it did not maintain normal rates of protein synthesis and mitosis in the denervated blastema (Rathbone et al., 1980). cGMP also shows elevated amounts during the early and late dedifferentiation stage, and it declines during the cone stage, similar to cAMP. The levels of cGMP matched those of the intact limb after the differentiation started (Liversage, Rathbone, and McLaughlin, 1977). This indicates that both cyclic nucleotides are involved in the control of cell proliferation.

Protein kinase C (PKC) is another enzyme implicated in signal transduction. Increase in PKC activity has been found in regenerates at 14 days after amputation in both the membranes and the cytosol, with partial translocation to the membranes after that period. Translocation of PKC is related to its activation. Denervation was shown to prevent this translocation only in the blastema mesenchyme. The process of limb regeneration or denervation did not stimulate PKC in the nervous system but did increase the translocation to the membranes in the spinal cord and the brain. These studies suggest that a nerve-derived mitogenic factor may work through the activation of PKC (Oudkhir et al., 1989). Protein kinase A, on the other hand, is decreased during the period of blastema accumulation, but it rises during the differentiation and morphogenesis stages (Laz and Sicard, 1982). Activity of two phosphoprotein phosphatates, the [Tyr] myelin basic protein (MBP) and the [Tyr] reduced carboxyamidomethylated lysozyme (RCML) has been also studied during newt limb regeneration. These two proteins represent substrates for two different types of mammalian phosphotyrosine phosphatases that are involved in signal transduction. With MBP, PPtase activity was elevated drastically after the mid- to late-bud stage and remained high up to the digital stage. With RCML, the activity fell at the mid- to late-bud stage and had slowly increased by the digital stage. This means that protein tyrosine phosphatase activity is lowest in the highly proliferative blastema (Sicard, 1993).

Calcium mobilization has also been found to affect cell division. Blastemata cultured in the presence of ionophore A23187 (transfers Ca^{2+} across the membranes) exhibited a twofold increase in the mitotic index. Such an effect is accompanied by elevation of cGMP and suppression of cAMP. Likewise, lowering of calcium intracellular levels resulted in suppression of mitosis (Globus, Vethamany-Globus, and Kesik, 1987).

6.3 Candidates for neurotrophic factors

Several criteria were outlined by Brockes (1984) for a molecule to serve as a neurotrophic factor: (1) It should be present in the blastema; (2) it should be reduced upon denervation; (3) it should be mitogenic to the blastema cells; (4) it should correct for denervation; and (5) its specific removal should block regeneration. Many substances have in the past been tested for their effects on blastema cells upon denervation. Even though many of them fulfill some of the criteria listed, we have not yet been able to identify *the* neurotrophic factor, nor do we know anything about its regulation in relation to regeneration and the regenerative ability of different amphibia.

6.3.1 Fibroblast growth factor (FGF)

FGF is a small polypeptide of about 13,000 Da that was isolated by Gospodarowicz (1974) and shown to be mitogenic in 3T3 cells. The role of FGF in limb blastemata was subsequently studied. Infused FGF in denervated newt blastemata restored mitotic activity (Mescher and Gospodarowicz, 1979). When FGF was added in blastema explants *in vitro*, it was able to promote ^{3}H-thymidine incorporation into DNA. In fact, the effect of a concentration of 10 ng/ml was equal to that of brain extracts. Furthermore, this dose was able to maintain activity of acetylcholinesterase in cultured newt muscle. FGF (Boilly et al., 1991) and FGF receptors (Poulin et al., 1993) are present in the newt blastema, but the exact role of FGFs in the neurotrophic stimulation of regeneration is not clear.

6.3.2 Glial growth factor (GGF)

GGF was identified by its activity on Schwann cells. These cells respond to extracts from the brain and pituitary. A basic protein of about 31,000 Da is responsible for this activity. GGF is present in the newt nervous system and also in the blastema. The levels of GGF decrease after denervation, and GGF increases mitogenic activity of the denervated blastema. When the 22/18 antibody is used (blastema cells whose division is dependent upon nerves are stained), a sevenfold increase in thymidine uptake was noticed (Figure 6.2). These data suggest that GGF fulfills the major necessary criteria for being a neurotrophic factor (Brockes and Kintner, 1986). It remains to be shown, however, that the specific removal of GGF will block regeneration.

6.3.3 Transferrin

Transferrin is a protein that is normally abundant in peripheral nerves and accumulates to even higher concentrations upon nerve regeneration. During limb regeneration, transferrin levels increase in the adult sciatic nerves in both axons and Schwann cells and is transferred by fast

Figure 6.2 Glial growth factor stimulates cell division of 22/18 positive cells. Blastema was isolated from denervated limbs and placed in culture in the presence of GGF. One day later the ^{3}H-thymidine was added to the cultures and the next day the explants were processed for staining and autoradiography. A: Immunofluorescence of a section stained with 22/18. B: The same section showing two thymidine-labeled nuclei, at least two of which are present in 22/18 positive cells (arrowheads). (Courtesy J. P. Brockes.)

anterograde axonal transport throughout all stages of limb regeneration. Transferrin seems to be secreted from the distal ends of the regenerating axons, and its level is dramatically decreased after axotomy (Mescher and Munaim, 1984; Kiffmeyer, Tomusk, and Mescher, 1991).

6.3.4 Neuropeptides

The possibility has also been investigated that some small neuropeptides are present in the limb and play roles in regeneration. One of them, Substance P, seems to satisfy some of the criteria for a neurotrophic factor. It is present in the epidermis, but not in the mesenchymal blastema cells. Higher reactivity was observed at the stage of rapid blastema growth. Substance P can also stimulate blastema cell proliferation at the very low concentration of 10 pg/ml. Finally, an antibody to this peptide was able to block the mitogenic effects of nerves on the regeneration blastema (Globus and Vetnamany-Globus, 1985; Globus and Alles, 1990). Another peptide found in the nervous tissues of the newt as well as in the regenerating blastema is the Hydra Head Activator, a molecule responsible for the regeneration of the head in Hydra (Fuentes et al., 1993). Neurotensin has also been reported to be present in the basal layers of the newt limb epidermis (Globus and Alles, 1990).

6.3.5 Nerve growth factor (NGF)

The effects of this factor on limb regeneration in the axolotl and frog tadpoles have been studied (Weis and Weis, 1970; Weis, 1971). NGF was shown to augment the size of the regenerates when compared to the untreated normal regenerates. The study did not examine the effects of exogenous NGF on denervated limbs; therefore, it is not clear whether NGF fulfills the criteria for a neurotrophic factor. However, NGF and other related molecules, the neurotrophins, are able to stimulate growth and survival of neurons, and they are also expressed in the developing rat limb bud (Henderson et al., 1993).

6.4 Addendum

6.4.1 Hormones and limb regeneration

Hormones are produced by several glands of the endocrine system, and they play crucial roles in sustaining normal metabolism and growth. It is therefore anticipated that several hormones are important for maintaining normal physiological events during the regenerative growth. Since the early 1900s, scientists have examined the effects of hormones by depleting the organism of specific glands. This does not necessarily point to the effect of a particular hormone, because some glands, such as the pituitary, are responsible for synthesizing more than one hormone. Therefore, the effects of gland removal cannot be directly correlated with the effect of a particular hormone; however, in some experiments where replacement was performed, some direct links can be established.

In 1910, Walter studied the effects of thyroidectomy on limb regeneration of the adult newt. The thyroid was removed at the same time that the limbs were amputated. The result was stumped or stunted malformed regenerates. Abnormal regeneration was also seen when the thyroid was removed 2 days prior to amputation (Richardson, 1940). Interference with the thyroid stimulating hormone (TSH) resulted in inhibition of tail regeneration. When thyroxin, the major hormone produced by the thyroid gland, was added to the water containing thyroidectomized newts, it did not support limb regeneration or survival of the newts. Similarly, induced hyperthyroidism was found to impede regeneration, indicating the importance of balanced physiological levels in growth (Liversage, McLaughlin, and McLaughlin, 1985).

Hypophysectomy severely affects the growth of the regenerate regardless of the time of removal. While this surgical procedure does not seem to affect the wound healing and early dedifferentiation phases, growth of the regenerate is halted (Hall and Schotte, 1951; Schotte and Hall, 1952). The role of adrenal hormones was studied by using the drug Amphenone B, which suppresses the synthesis of adrenal cortical hormones (as well as thyroid hormones). Regeneration in animals receiving such a treatment was thwarted; it was restored in hypophysectomized newts when cortisone was injected together with Amphenone B, but not when adrenocorticotropic hormone (ACTH) was injected with the drug (Liversage et al., 1985). In other studies, insulin has been implicated in limb regeneration. By removal of the pancreas it was shown that limb regeneration can be inhibited (Vethamany-Globus and Liversage, 1973; Vethamany-Globus et al., 1984).

As stressed previously, studies in which removal of the whole gland was performed cannot conclusively target the role of individual hormones. Some insights on this problem were obtained by the replacement experiments, in which individual hormones or mixes of some were administered to the animal after removal of the gland. For example, better regeneration was obtained when extracts of growth hormone (GH), TSH, and gonadotropin were injected into hypophysectomized newts than in untreated hypophysectomized and thyroidectomized newts. However, when thyroxine was included in the mix, normal regeneration resulted in both groups. This experiment stresses the importance of thyroxine alone. Similarly, ACTH resulted in good survival and regeneration of hypophysectomized newts, indicating cooperation between the pituitary and adrenal glands (Liversage et al., 1985).

Prolactin and growth hormone, both produced by the pituitary, have been implicated in regeneration as well (Wilkerson, 1963). Both are imperative for differentiation and growth of regenerating limbs, and both induce regeneration when added exogenously to hypophysectomized newts (Tassava, 1969). The combination of both prolactin and thyroxine was especially effective. While all these experiments were very informative and well designed to decipher the roles of the endocrine system in limb regeneration, the regulation and mode of action of individual hormones is far from realized in the field of regeneration. This results from the limitations of such surgical procedures, as already mentioned, and the impurity of preparation used in many experiments.

The mode of action of hormones in general has recently been studied in great detail at the molecular level in other systems. We now know that some of these hormones act through the generation of second messengers and calcium mobilization, thus leading to cellular responses and cell proliferation. Other hormones are the ligands for nuclear receptors, which belong to the steroid and thyroid superfamily (Evans, 1988). This superfamily embraces members such as the receptors for estrogen, glucocorticoid, androgen, vitamin A, and vitamin D, to mention a few (see also Chapter 17). These nuclear receptors are transcriptional factors; once activated by the binding of the ligand, they are responsible for the activation of transcription of numerous genes leading to a certain state of metabolism and physiology. Some of the hormones, such as insulin and thyroxine, shown to be effective on limb regeneration, have been shown to promote cell proliferation and DNA synthesis in newt blastemata. Molecular studies with these receptors are nonexistent. It is conceivable that studies of this kind can shed light on the mode of action and regulation of hormones during limb regeneration; detailed studies will establish a more concrete and accurate picture of the action of hormones during limb regeneration.

6.4.2 Role of the immune system

The involvement of the immune system in the regeneration process is obscure. No clear or definite immune response has been linked to the process. Indirect evidence, however, implicates compounds or molecules of the immune system. Intriguingly enough, when agents that either stimulate or suppress the immune system were employed to explore their effect on limb regeneration, the end result was always inhibition or abnormal regeneration. These include the immune suppressants cyclophosphamide; cyclosporin and antilymphocyte serum; thymic extract fractions that induce T cells; the lymphokines; interleukin 1, 2, and 3; the cytokines interferon beta and tuftsin; and the cobra venom factor, which is a complement activator (Sicard, 1989). Cyclophosphamide treatment had an adverse effect on frog limb regeneration depending on the time of treatment. While treatment before amputation inhibited regeneration, concomitant treatment with amputation resulted in enhancement of regeneration. Challenge and stimulation of the immune system with skin allografts inhibited or enhanced regeneration, depending on the time between the challenge and the amputation. Challenge 2 weeks prior to amputation favored blastema growth, whereas challenge 1 week before, or concomitant with, amputation did not. This means that an active cellular immune response retards the early events of regeneration. This could imply that down-regulation of the immune system is imperative for the beginning of regeneration. The cyclophosphamide treatment also confirms this, since accelerated regeneration is observed when it was administered at the time of amputation (Michael, Aziz, and Fahmy, 1993).

Molecular data in this area are virtually nonexistent. The availability of axolotl complement factor C3 antibodies makes it possible to ascertain the expression of this factor during limb regeneration. The factor is very strongly expressed in the blastema and the differentiating muscle, but it is virtually absent from the mesodermal tissues of the developing and intact limb (Tsonis and Lambris, unpublished observations).

Is the newt immune system a Pandora's box for the field of limb regeneration? Down-regulation of the immune system during blastema formation would mean that the organism cannot very effectively protect against "foreign" antigens at the time of regeneration. Can the reexpressed embryonic antigens found in blastema cells be considered as "foreign," that is, something analogous to cancer cells? If that is the case, then down-regulation is imperative to ensure that subsequent regeneration will not be fought by antibodies or cells of the immune system. Can this explain the apparent lack of tumors in amphibia with regenerative capacity? If such a mechanism exists – by far very unique to amphibia – it could provide fundamental information about regulation of the immune system and help us fathom the basic physiological mechanisms underlying limb regeneration.

7

Protein Synthesis in the Blastema

The dynamic process of cell dedifferentiation, and the subsequent proliferation of the regenerative blastema, entail specific gene activity and protein synthesis. Detailed analysis of protein synthesis has been achieved by the use of two-dimensional polyacrylamide gel electrophoresis. This technique, unlike the regular one-dimensional gel electrophoresis, provides a better picture of protein synthesis, with a resolution of hundreds of proteins. Such an analysis has provided interesting insights on patterns of protein expression during blastema formation. Slack (1982) performed such experiments where proteins were analyzed and compared from different stages of regeneration-specific tissues (mesenchyme versus wound epithelium) and the embryonic limb. These studies showed that protein synthesis in the regenerating blastema is very similar to that of the embryonic limb and pinpointed differences specific to the mesenchyme or the wound epithelium. Although those studies were able to establish a qualitative picture of protein synthesis, they lacked any quantitative analysis in comparison with the intact limb.

Detailed comparison with the intact limb was performed by Tsonis and colleagues (Tsonis, Mescher, and Del Rio-Tsonis, 1992a, 1992b; Tsonis, 1993). Using a powerful program in which the gels or the autoradiograms were digitized, images from different stages of the blastema and the intact limb were superimposed upon each other, thus creating a protein database that then allowed quantification and analysis of the expression of each protein. With this procedure the expression of about 800 proteins in the intact and regenerating limb was examined, and the expression of each protein was then compared at different stages. The overall comparison showed that protein synthesis 1 week after amputation, that is, in the dedifferentiation stage, was severely suppressed. In addition, it was demonstrated that protein synthesis differs considerably in 1- and 2-week regenerates (Figure 7.1). Blastema formation was marked by the expression of 134 proteins unique to this stage, whereas only 26 proteins were unique to the dedifferentiation stage (Figure 7.2). This protein database not only provided a comprehensive picture of protein synthesis in general, but it could also be useful to isolate and characterize proteins that show interesting patterns of expression.

To concentrate further on proteins that seem to be important for regenerative processes, this database was compared with protein synthesis profiles from the regenerating tail and the regenerating lens. All these regenerative processes undergo dedifferentiation that necessitates rigorous remodelling of the extracellular matrix (Chapter 4). It is possible that regulation of unique molecules happens in all three tissues. Therefore, comparison of protein synthesis in all three tissues could lead to the identification of crucial factors. Indeed, the comparison showed that there are common patterns of protein synthesis among regenerating limb, tail, and lens. In fact, one protein seems to disappear from all three tissues; others are expressed uniquely in all, or disappear in limb and tail but not lens. Sequencing of the protein that disappears in all three tissues has shown similarities with a proteoglycan (Tsonis and Del Rio-Tsonis, 1995), whereas sequencing of a protein that is up-regulated in the regenerating tissues has identified a keratin (Tsonis, Mescher, and Del-Rio Tsonis, 1992). These results indicate the usefulness of such an approach in characterizing factors that might initiate regeneration and might provide leads and tools for further work on regulation.

Similar studies have indicated that there are differences in protein

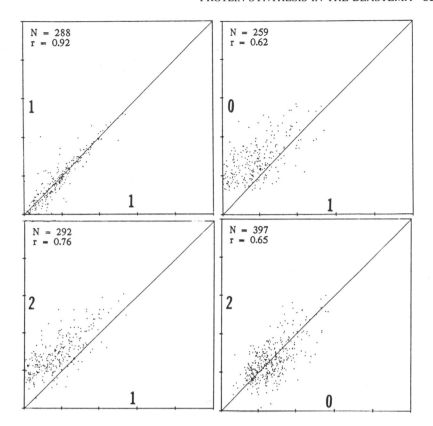

Figure 7.1 Statistical analysis and presentation of protein synthesis during limb regeneration in *Notophthalmus viridescens*. Comparisons of gels from intact (0), 1-week regenerating (1), and 2-week regenerating (2) limbs. The more similar protein patterns are on a two-dimensional gel, the closer the matching spots (of number *n*) (representing common proteins) are to a diagonal line, and the correlation coefficient (*r*) is closer to 1. In the upper left panel, comparison of two different runs of the same sample (1-week regenerating limb) indicates a correlation coefficient of 0.92, which in turn means very similar, if not identical, patterns. However, when runs from intact (0), 1-week (1), or 2-week regenerating limb (2) are compared to each other, it can be observed that protein synthesis patterns are not similar. (Tsonis, Mescher, and Del-Rio Tsonis, 1992.)

synthesis along the proximal–distal axis. These studies examined protein synthesis in proximal and distal blastemata before and after treatment with retinoic acid. Even though characterization of proteins by sequencing was not attempted, such analysis can provide information on molecules involved in pattern formation during regeneration (Carlone and Boulianne, 1991).

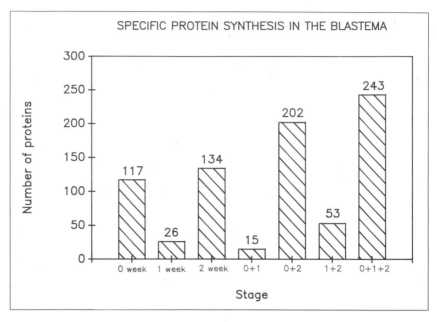

Figure 7.2 Specific protein synthesis in different stages of limb regeneration. 0: intact limb, 1: 1-week blastema, 2: 2-week blastema; 0+1, 0+2, 1+2, 1+2+0 indicate combination between the different stages. The numbers over the bars indicate the number of proteins found to be expressed only in the particular stage, and they were derived from the quantitation of all proteins in gels from different stages. (Tsonis, Mescher, and Del-Rio Tsonis, 1992.)

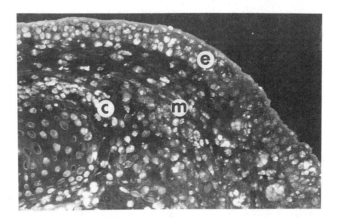

Figure 7.3 Localization of chromosome breaks in a section through the blastema 3 weeks post-amputation. Nuclear staining (depicting active nick translation and therefore breaks) can be seen in all tissues of the blastema including epithelium (e), the mesenchyme (m), and the differentiating carti-lage (c).

The active synthesis of new proteins during blastema formation is also reflected in active DNA transcription during this stage. Active gene transcription can be measured by means of single-strand nicks and repair in DNA. Actively transcribed genes are subjected to augmented strand repair. Such repair in the chromosomal breaks can be quantitatively measured by the activity of the enzyme ADP-ribosyl transferase, which catalyzes the transfer of ADP-ribose onto histones and is entirely dependent on DNA containing single-strand nicks (Benjamin and Gill, 1980; Farzaneh et al., 1982). Such activity has been measured during limb regeneration, and it was found to increase twenty- to fiftyfold during blastema formation (Tsonis, unpublished). The nicks that represent active transcription can be seen in Figure 7.3, which is the result of *in situ* nick translation with biotin-11-UTP and immunofluorescence with an anti-biotin antibody in a 3-week blastema section.

Limb Cells
in Vitro

Studies that deal with differentiation can benefit enormously from *in-vitro* systems. Cell behavior can be studied more readily with cells grown in culture. For example, the transdifferentiation of the pigmented epithelium of the newt eye into lens (Eguchi, 1988) has been studied in clonal cell cultures, providing unequivocal proof of the process of dedifferentiation and metaplasia. Furthermore, cultured cells can be treated with factors and their response more effectively studied. The 101/2 3T3 cells can select a differentiation pathway to myocytes if treated with 5-azacytidine (Taylor and Jones, 1979). This approach led to the identification of Myo-D, a key regulatory factor in myogenesis (Lassar, Paterson, and Weintraub, 1986). Availability of *in vitro* systems for the blastema would be of enormous interest for the study of dedifferentiation and subsequent differentiation. In addition, manipulation of blastema cells can be possible, which could allow transplantation experiments with cells possessing markers.

Historically, blastema cells have proven difficult to culture. When dissociated, the cells did not usually grow well (compared with other newt cells, such as the pigment epithelium). In addition, growth was always disturbed by infections from bacteria lurking in the newt limb. A well-balanced culture solution also seems to be necessary for good growth. Fimian (1959) reported that blastema cells in a simple balanced solution with glucose could grow, as seen by the mitotic index, but could not differentiate. Jabaily, Blue, and Singer (1982) found that blastema cells could grow and differentiate into myotubes if dissociated mechanically. The medium that seemed to be the best was the CMRL 1415 and, later, Leibovitz-15 (L-15), containing galactose. Minimal essential medium with Eagle's balanced salt solutions or combinations with L-15 can be used. The osmotic pressure should be adjusted appropriately for salamanders, and the cells should grow at a pCO_2 concentration of 2 percent at 25°C. A pH of 7.2–7.4 is maintained by the amount of bicarbonate. Antibiotics, such as penicillin/streptomycin, and antifungal agents are imperative to avoid contaminations often observed in such cultures. Despite all these efforts, however, it was not until the late 1980s that culture of newt limb cells was achieved. A very effective method was to place an explant of blastema in culture (Hinterberger and Cameron, 1983). Blastema cells can grow out of the explant very well. Similarly, muscle explants give rise to well-growing myoblasts that can, in fact, fuse into myotubes. These pleiomorphic cells express myf-5 (Tsonis, Washabaugh, and Del Rio-Tsonis, 1995).

Culture of blastema explants has revealed that there are at least four different types of blastema cells classified according to their morphology (Ferretti and Brockes, 1988). The fibroblast-like pleiomorphic cells look very much like the myoblasts coming out of a muscle explant, the bipolar cells, and the giant multinucleated cells (Figure 8.1). It is possible that these cells are derived from muscle, nerves, and bone, respectively. The fourth cell type, the so-called signet cell (Maier and Miller, 1992), resembles the osteoprogenitor cells or the cells that fuse with the multinucleated cells. In fact, the signet cells are monocytes that fuse to form osteoclasts. When newts were injected with ^{3}H-thymidine before amputation (so that only the cells of blood origin are labeled) and the blastema was explanted *in vitro* 2 weeks later, the label was present only in the signet cells and the osteoclasts (Figure 8.2) (Washabaugh and Tsonis, 1994). The main conclusion from these studies is that the blastema is not composed of a homogenous cell population, which is implied by histological studies. The pleiomorphic, bipolar, and signet

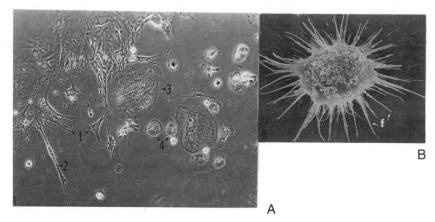

Figure 8.1 A: The different cell types growing out of a 2-week blastema explant placed in culture. Note the pleiomorphic cells (1), the bipolar cells (2), the giant cells (3), and the signet cells (4). B: Scanning electron micrograph of a pleiomorphic cell indicating the numerous filopodia (f) and ruffled surface.

cells express the regeneration-associated antigen that interacts with the 22/18 antibody, and so do the cells coming out from the muscle explants (Ferretti and Brockes, 1988). From this we can in fact concentrate on cells from the muscle explants, since they grow much better than cells from the blastema explants. If these cells are allowed to grow to confluency, they fuse and form myotubes. Such a property is very significant for the study of dedifferentiation and differentiation processes. Cartilage differentiation has not been observed routinely (Hinterberger and Cameron, 1991). There might be some other requirement missing from the culture medium that could promote differentiation to chondrocytes; with the culture system now available, such studies can be pursued.

The different cell types growing out of a blastema explant seem to communicate extensively. As mentioned, fusion of the monocytes forms osteoclasts, while fusion of the pleiomorphic cells results in the formation of myotubes. Also, when blastema mesenchyme is placed in culture with epithelial cells, they form two distinct layers and in fact grow very well. This could be from the growth-promoting activity that the wound epithelium exerts upon the blastema.

With limb cells compatible for regeneration grown effectively in culture, more experiments can be designed to fathom some of the mech-

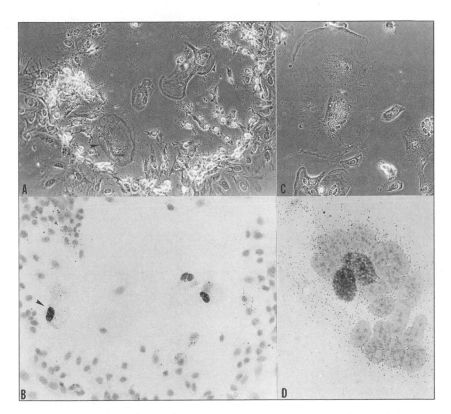

Figure 8.2 The signet cells are mononuclear leukocytes that fuse and produce the giant cells, the osteoclasts. In this experiment ^{3}H-thymidine was injected in newts the day before limb amputation. This procedure will allow only the labeling of blood cells and wound epithelium. The blastema is then placed in culture 2 weeks later. Note that only the signet cells are labeled, whereas all the pleiomorphic cells are not (B, arrowhead). In D the giant cell in C with two out of its 20 nuclei labeled is shown. This indicates that two signet cells have been fused in to form the multinucleated osteoclasts.

anisms of limb regeneration. These cells have been used to study de-differentiation (see Chapter 4) and also in studies dealing with retinoic-acid induction of gene expression. These cells seem capable of being transiently transfected by genes, but the stable integration of genes could be a problem. If this problem can be circumvented, the different cell types of the blastema cells can be expanded, perhaps by the integration of an oncogene, and different cell lines of blastema cells can be obtained. Such cell lines could prove vital for studying cell regulation

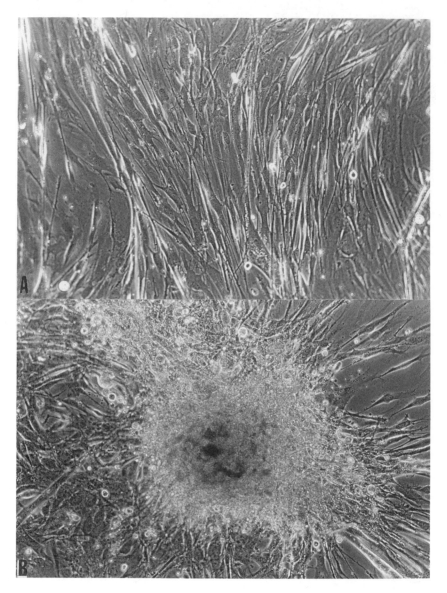

Figure 8.3 Immortalization of pleiomorphic cells derived from muscle explant. A: Normal pleiomorphic cells that have been grown in culture for 2 weeks. Note that these cells have started aligning to form myotubes. B: Pleiomorphic cells transfected with the large T antigen–containing plasmid. Cells fail to differentiate and pile up, exhibiting a transformed phenotype.

and differentiation and could lead to the identification of crucial factors in regeneration.

Immortalization of limb cells has been attempted by using the large T antigen. The initial experiments involved cells from muscle explants. Successfully transfected cells can resist selection (the antibiotic G418) and grow in piles characteristic of transformed cells (Tsonis et al., 1995). They differ dramatically from the untransfected ones, and they do not fuse to form myotubes (Figure 8.3). Both cell types, however, express myf-5 (Figure 8.4). So far, these experiments have shown that transformation is feasible. Both normal and transformed cells were also found to participate in regeneration upon transplantation into the amputated limb. For such studies, cells were labeled with a fluorescent dye that is integrated in the cell membrane and does not diffuse to other cells. Detection of this fluorescent marker indicated participation of these cells through 30 days of regeneration. Furthermore, the marker was present in regenerating muscle as well as in regenerating cartilage (Tsonis, Washabaugh, and Del Rio-Tsonis, 1995) (Figure 8.5). These results show convincingly that dissociated cells can participate in regeneration and that they can differentiate into different cell types as well. Such a participation of cultured myotubes has been also shown by Lo et al. (1993). If these cells are clonal and, as believed, the product of muscle dedifferentiation (they express myf-5), these results also show the ability of these cells for transdifferentiation. Stable integration of a plasmid into the genome has proven difficult to achieve. In fact these cells possess the plasmid episomally (Tsonis et al., 1996). This is advantageous to the system, because the reporter gene is expressed for long periods and can be traced upon transplantation. On the other hand, stable integration could help in producing stable transfectants that could lead the way to the generation of transgenic limbs or animals. This problem could be circumvented by the use of retroviral vectors. Such vectors, however, are restricted to some hosts only, especially the ones derived from Moloney murine leukemia virus. Recently, a pseudotyped retroviral vector (with the genome of one virus encapsidated by the envelope protein of another) using the vesicular stomititis virus G glycoprotein has been found to infect newt limb cells (Burns et al., 1993, 1994) (see Part III). Such viruses can be used to perform a stable transfer of a gene into cultured cells or *in vivo*. In fact, these viruses were tested in newt cell lines, such as the ones mentioned previously, and

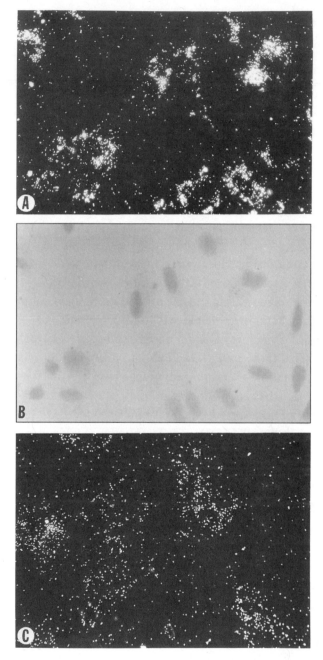

Figure 8.4 A: Expression of myf-5 in immortalized cells. B: Phase contrast image of cells in A counterstained for the nuclei. C: Expression of myf-5 in normal untransfected cells. Note the higher level of expression in the immortalized cells.

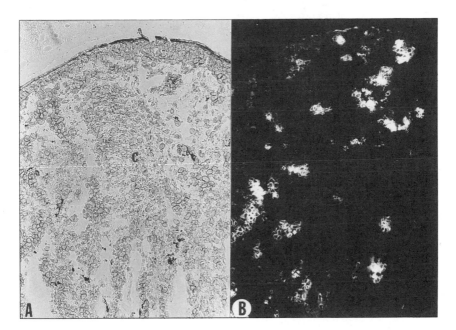

Figure 8.5 Participation of the cultured blastema cells in normal regeneration. The cells were treated with DiI, a fluorescent dye, dissociated and transplanted into an amputated limb. One month later, sections of the regenerate were taken and observed under fluorescence. Note labeled cells in the muscle and the cartilage (c) of the regenerate.

were shown to infect, integrate, and express the reporter gene. This development opens new avenues in the study of limb regeneration (see Part III). Newt cells can now be infected very efficiently *in vivo*, and several aspects of dedifferentiation, differentiation, and positional identity can be studied.

9

Tissue versus Epimorphic Regeneration

Now that general descriptions of the events and stages of regeneration have been presented, we should consider the differences between true epimorphic regeneration, as described for the salamanders, and tissue regeneration, which might underlie the events that we see with the limited regeneration in frogs and other vertebrates. The term "epimorphic regeneration" was coined by Morgan (1901) to characterize the type of regeneration seen in the replacement of whole body parts, such as the limbs, where dedifferentiation and redifferentiation of various cell types occur.

The anura amphibia can regenerate their limbs rather well before metamorphosis approaches. During metamorphosis the regenerative ability declines. Thus the regenerative ability is lost when amputation is performed at proximal levels but not at distal levels. The ability of regeneration of the distal levels is lost thereafter. But some anura can still, even after metamorphosis, regenerate hypomorphic structures or elongated spikes that look like extensions of the extremities. One such anuran is the South African clawed frog, *Xenopus laevis* (Dent, 1962). Skowron has studied the histological features of such regenerates. He observed that the hypomorphic regenerates are developed from the end of the removed bone and that they consist of cartilage and skin without a pattern (Figure 9.1). Muscle or joints are not formed. The regenerated tissues are well vascularized and innervated, but blood vessels and nerves are not arranged as they would be in a normal limb (Skowron and Komala, 1957; Skowron and Roguski, 1958).

Different ideas have been presented in the literature to account for the loss of the regenerative ability in frogs and higher vertebrates. The process of metamorphosis depends on thyroxine, but it is doubtful that this hormone by itself would inhibit regeneration. Thyroxine alone does not affect normal regeneration (Hay, 1956), and it does not support regeneration in thyroidectomized animals. It has recently been implicated in patterning of the regenerating limb through similar and competitive pathways to retinoic acid in axolotls (Vincenti and Crawford, 1993). Singer (1951) performed an experiment that suggested frogs have an inadequate nerve supply in the limbs for regeneration. He rerouted nerves from the hindlimb to the amputated forelimb, and this augmentation of the nerve supply was sufficient to induce regeneration

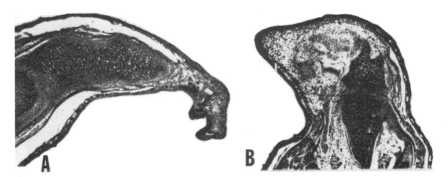

Figure 9.1 Sections through amputated frog (*Xenopus laevis*) limbs. A: Hypomorphic regenerate 32 days after amputation at stage 57. The end of the femur (amputation site) is at the left and is surrounded by a ring of cartilage that is continuous with the cartilage of the regenerate. B: Hypomorphic regenerate 32 days after amputation at stage 60. Bone has been formed around the cartilaginous ring that is surrounding the end of the femur (amputation site). (After Dent, 1962.)

(see Chapter 6). Thus, in anura, unlike the urodeles, there seems to be a correlation between innervation and ability of regeneration (Scadding, 1982). It could be that the thyroxine-induced metamorphic changes evoke physiological modifications affecting the release and/or function of the necessary neurotrophic factors. A different idea has been offered by Scadding and Maden (1994), who have found that *Xenopus* limbs do not organize gradients of endogenous retinoic acid in the same way as urodeles. Given the action of retinoic acid on limb morphogenesis during regeneration, the authors imply that its absence in the frog leads to the generation of disorganized and hypomorphic structures.

Evidence has also been presented on the intrinsic nature of the regenerative ability in frogs. Muneoka, Holler-Dinsmore, and Bryant (1986b) studied regeneration in *Xenopus laevis* at different developmental stages and after removal of different portions of tissues. As development progressed from stage 53 to 59, it was found that as more tissue was removed, regeneration was more likely to be possible. For example, at stage 53 there was good regeneration ability after removal of one or three digits, but by stage 58 or 59 regeneration could be obtained only when three digits were removed. These results indicate that intrinsic factors connected to growth during regeneration might be involved in the regenerative loss in frogs. This issue was further pursued in a different series of experiments. Normally regenerating limb buds of stage 52–53 were grafted onto the limb stumps of postmetamorphic frogs, which normally do not regenerate their limbs. The limb buds became well vascularized and innervated and, when amputated, regenerated almost normal limbs. The opposite experiment resulted in spike-like regeneration characteristic of postmetamorphic frog limb regeneration (Sessions and Bryant, 1988). These results strengthen the idea that the *Xenopus* host environment is permissive for regeneration and that the ability for limb regeneration is an intrinsic property of the young tadpole limb cells, which is lost during ontogenesis.

Is regeneration of hypomorphic structures epimorphic? This is one question that several scientists have attempted to answer over the years. Epimorphic regeneration is defined by formation of a blastema. So far, histological studies indicate that during the hypomorphic regeneration in frog limbs, cartilage and skin are produced by a direct proliferation event, rather than through dedifferentiation and formation of a blastema. Goss and Holt (1992) took another approach. When a blastema from a newt (epimorphic regeneration) is stripped of its wound epithelium and placed in the body cavity, so that the generation of another wound epithelium is prevented, subsequent regeneration is prevented. This has been interpreted to mean that a true blastema fails to regenerate because of the loss of the appropriate signals from the wound epithelium. Goss tested this concept with the frog limb and likewise found that the amputated limb was unable to form even a hypomorphic structure when inserted into the body cavity. Goss argues that if a blastema is not formed and that if frog limb regeneration is the result of direct proliferation, then hypomorphic regeneration should not have been inhibited. This is taken as a sign of true blastema involvement in *Xenopus*. Such an experiment, however, cannot rule out the possibility that proliferation directly from the cut bone is also inhibited by the absence of wound epithelium. Wound epithelium is known to pass proliferative signals to the underlying mesenchyme with the synthesis of specific molecules such as FGFs (Boilly et al., 1991; Niswander et al., 1993).

The need for molecular markers is urgent in order to solve such problems. If markers for blastema cells are generated when regeneration is initiated (early stages), these markers can then be used to examine the possibility of blastema presence during the regeneration of the hypomorphic regenerates in the postmetamorphic frog. Recently Ide and coworkers (1994) have taken a different approach to analyze the regenerative potency of *Xenopus* limbs. They established cultures of limb-bud cells and blastema cells from limbs taken at different stages of development. The stages ranged from 51 to 57. Regenerative capacity of *Xenopus* limbs is gradually lost after stage 55. When cultures of limb-bud cells at high cell density were performed, it was observed that the limb-bud cells differentiated to cartilage. Many nodules were observed at early stages (51–53), while few or no nodules were seen at stages 55 and 57, respectively. The capacity for chondrogenesis of the mesenchymal cells was gradually lost as regenerative ability was lost. The same was the case when wrist-level blastema cells from all these stages were placed in culture. Cartilage nodules were present when stage 51–53 blastema cells were cultured, but they were absent when stage 57 blastema cells were cultured. These results implicate the cartilage differentiation process in the loss of the regenerative ability. Loss of cartilage differentiation ability is accompanied by loss in the regenerative ability of the animal. This ability of limb-bud mesenchymal cells to undergo spontaneous chondrogenesis when plated at high cell density is well established when cultures of these cells are used from chick wing buds or mouse limb buds (Ahrens, Solursh, and Reiter, 1977). The problem here is that the ability for chondrogenesis in chick and mouse limb buds is very good, but the regeneration potency is not, unless the mesenchyme (in chick) is provided with AER signals (Hayamizu et al., 1994; Taylor et al., 1994). Such experiments could shed light on this interesting correlation between *in vitro* chondrogenesis and regenerative capacity in vertebrates.

On the other hand, however, the blastema cells isolated from the newt and placed in culture do not readily differentiate to chondrocytes. Myogenesis is the most common event seen in these cultures from the blastema or muscle explants. It will be of interest to observe chondrogenesis in these cultures as well. According to the previously mentioned correlation these cells from the newt limb should be differentiating to chondrocytes, because the regenerative capacity of these limbs is very good. Such experiments are now feasible with the existence of cell lines from the newt limb, and they should provide insights into the cellular mechanisms underlying the loss of the regenerative capacity in post-metamorphic anurans.

10

Bioelectricity and Limb Regeneration

The idea that electric fields are associated with living tissues is in fact very old, stemming from observations of ancient Egyptians and Greeks (Vanable, 1991). The existence of "animal electricity," however, was shown in the eighteenth century by the experiments of Galvani and Volta. Those experiments observed twitching of the frogs' muscle, which was later attributed to the nerves. In the first half of the nineteenth century, a seminal work by Du Bois-Reymond provided the first evidence that a current of about 1 microampere was flowing out of the skin of a wounded finger. Following this realization, currents were associated with wound healing and development. This also provoked interest in the involvement of currents in regeneration.

In 1941, Monroy first reported that an electric potential exists in the amputated salamander limb by showing its existence between the surface of an amputated stump and the skin of the stump. Smith (1974) suggested that currents passing through the stump of an amputated frog limb were responsible for stimulating limb regeneration. This experiment sparked the search for these currents during limb regeneration and for their effects upon the process. Borgens, Vanable, and Jaffe (1977a, 1977b) attacked this problem by using a very sensitive vibrating probe; their results showed that large currents leave the stumps of regenerating newt limbs. These currents of 10–100 $\mu A/cm^2$, leave the stump during the first 5–10 days after amputation. Other currents, of 1–3 $\mu A/cm^2$ enter the limb through the skin and are dependent on entry of sodium ions into the skin (Figure 10.1). Upon denervation of the limbs there was no effect in the current surface density. In fact, the stumps of the denervated limbs generated larger currents. This means that the motor and sensory nerves do not contribute to the generation of the currents.

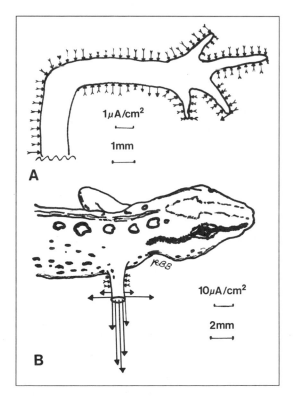

Figure 10.1 Pattern of currents around an intact newt forelimb (A) and forelimb stump after amputation (B). Arrows indicate direction of the currents and magnitude of current density as measured 0.3 mm from the surface. The difference in scale in A and B facilitates the visualization of small currents of intact skin. (Courtesy R. B. Borgens.)

More detailed experiments, in which the sodium concentration was manipulated in the medium in which the animals were placed, provided direct support that the regeneration currents are skin-driven. High concentrations of sodium resulted in increase of the currents, while lower ones resulted in decrease. When amiloride, a substance that blocks sodium entry into the skin, was used, the peak current was reduced dramatically within 15 minutes. By 50 minutes the current had fallen by 90 percent. Regeneration currents were found not to be unique to the newt *Notophthalmus viridescens*. Similar currents were observed in other salamanders and newts, such as the ones that belong to the Plethodontidae, Salamandridae, Amblystomidae, and Proteidae (Borgens et al., 1984).

The initial studies assumed that the externally measured currents reflect the presence of internal currents that are capable of exerting a biological effect. Subsequent studies in fact showed that the electrical current in newt limb stumps is indeed internal (McGinnis and Vanable, 1986a, 1986b). When proximal fields were measured from the shoulder to the last 0.5 mm of the stump, they averaged 6.6 mV/mm immediately after amputation but fell to 0.6 mV/mm in 24 hours. When distal fields were measured (those in the distal 0.25 mm), they averaged 50 mV/mm at the time of amputation. The currents fell to 15 mV/mm in 6 hours and to 5 mV/mm by 24 hours. The distal field strength is within the range that is known to affect cell behavior *in vitro*. Blocking of sodium channels of the epidermis abolished the generation of the fields.

The apparent decrease of the currents within a few days after amputation poses the question of whether it occurs because of changes of the skin battery or because of the increase in the resistance of the stump current circuit. Measurements indicated that the wound epithelium showed increased resistance correlated with the decrease in the stump current. This suggests a role of the wound epithelium in controlling the intensity of the current that flows out of the regenerating limb. All these observations supported the notion that electric currents have a role in limb regeneration. The mode of action of such currents is not completely clear, but it has been suggested that their role in regeneration could be mediated by promoting nerve growth into the limb (Hearson, Eltinge, and Vanable, 1988).

More direct evidence for the role of electric currents in limb regeneration is provided by the effects of inhibition or augmentation of the currents. When the sodium-dependent currents were reduced by amiloride, regeneration of the newt limbs was blocked or deficient in 50 percent of the cases (Borgens, Vanable, and Jaffe, 1979a). In other experiments it was demonstrated that application of small currents (0.2 µA) could initiate limb regeneration in the frog *Rana pipiens* and form nerve trunks within cartilage (Borgens, Vanable, and Jaffe, 1979b). Regeneration was indicated by newly formed muscle ligaments and cartilage. In addition, it was observed that large amounts of nerve developed in the regenerated tissues. Similar experiments in *Xenopus laevis* confirmed the previous observations. It was noted that the stimulated regenerates had fortyfold greater nerve supply than the controls. These experiments suggest that the currents affect regeneration by stimulating nerve growth. This is reminiscent of the fact that frog limb regeneration can be initiated when increased nerve supply is introduced to the amputated limb (Singer, 1951). Similar experiments have noticed the beneficial effects of currents in rat limb regeneration (Becker, 1972). Lassalle (1980) proposed an argument, however, on the necessity of surface potentials for regeneration. Lassalle pointed out that variations in surface potentials during limb regeneration are identical with the ones produced when skin wound healing occurs alone, and, therefore, are not likely to control limb regeneration.

11

Postembryonic Induction in Amphibian Limbs

The process of limb regeneration in salamanders is fascinating and has no parallels among higher vertebrates. The mechanisms of regeneration have not been unraveled despite a deep understanding of the events that take place. For example, scientists discriminate between amputation and subsequent limb regeneration and mere injury of the limb. In other words, it is established that for regeneration to take place a simple injury is not sufficient. The aspect of an amputation has also been applied in the mechanisms of patterning during regeneration (see Part II). However, some early work has indicated that an injury is sometimes sufficient to induce limb formation. Della Valle (1913) reported limb induction by trauma. Multiple formation of limbs by trauma or by simple incisions made during limb development was the subject of numerous studies in the early 1900s (Korschelt, 1931). Such supernumerary limb formation adds to the mystery of regeneration. In fact, it seems that supernumerary limb regeneration is initiated by events similar to those in the case of amputation – that is, dedifferentiation and blastema formation. The big question, of course, is determining the limit of limb-inducing injury. What can we learn about this induction to help us understand the event of limb regeneration and apply this knowledge to other vertebrates?

In the 1930s, experimental work (chiefly from the Nassonov laboratory) demonstrated the amazing capability of limb tissues to induce limb formation postembryonically in adult amphibia. One of the most intriguing aspects of induction was that even a tight ligature of an intact extremity would induce the development of an additional limb, probably due to the disintegration of tissues resulting from such treatment. Nassonov experimented, therefore, with the ability of several tissues to induce limb development if placed underneath the skin of an intact limb. He found that transplantation of regenerating buds or pieces from other tissues was able to induce supernumerary limbs in axolotls. He concluded that cartilage possessed the best activity and gills the least (Nassonov, 1936, 1938, 1941). With the limited means of that era, Nassonov continued his experimentation and showed that hydrolysates of the tissues were able to induce limbs with two or three digits. Continuation of this work refined the requirement of the hydrolysates. Hydrolysis of cartilage with 25 percent formic acid and 1 percent HCl produced the most spectacular results; four- to five-digit legs were obtained. In fact, in many cases, two or three such legs could be formed at the same site. The biochemical action of these lysates is not understood, but it is conceivable that they contain polypetides beginning with tripeptides (Fedotov, 1946).

Such induction has also been studied by Carlson (1971) by implantation of *Rana pipiens* tissues into intact *Triturus* limbs. Carlson conclusively showed that the different tissues of the frog had varying supernumerary limb–inducing capacity. Lung tissue was the best, showing induction in almost 96 percent of the cases; urinary bladder and kidney tissue were also good and showed induction in 93 percent and 87 percent of the cases, respectively. Tissues such as stomach, colon, fat body, and phalanges were also good inducers, whereas brain, skeletal and cardiac muscle, skin, peripheral nerves, and liver were not.

Similar inductions were obtained by kidney and liver implants taken from *Taricha granulosa*, *Triturus viridescens*, and *Rana pipiens* when they were placed subcutaneously into limbs of *Triturus viridescens* (Ruben and Stevens, 1963). In other studies, however, tumor induction was observed in the limbs receiving normal kidney implants in *Xenopus laevis* (Balls and Ruben, 1964).

Implantation of tissues is not the only way by which supernumerary limb induction can occur. Butler and Blum (1955, 1963) reported that UV-irradiation could result in accessory limb production at the point of exposure. The authors explained their results by the induction of blastema cell proliferation due to the irradiation. Another spectacular method of inducing accessory limbs was using carcinogens to treat intact limbs . Breedis (1952) reported that supernumerary limb production was readily achieved when carcinogens were applied underneath the skin of the newt. The amount of carcinogens that Breedis applied was large and able to provoke widespread destruction in the limbs. Such destruction could have resulted in deviation of nerves, another method whereby supernumerary limbs can be produced. However, carcinogens induced accessory limbs even when low doses were applied and when toxicity did not disturb the integrity of tissues (Figure 11.1) (Tsonis and Eguchi, 1983). Carcinogens became probes in the study of limb induction and limb regeneration mainly due to their cancer-inducing property. In all subsequent studies, carcinogens were found to be teratogenic when applied to the regenerating limb, adversely affecting pattern formation (Figure 11.2). Interestingly, some of the abnormalities seen in the skeletal patterns resemble the ones induced by mutated Hox genes in

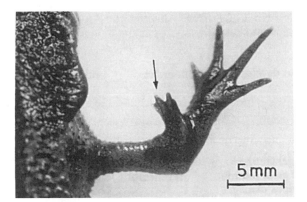

Figure 11.1 Induction of supernumerary limb formation in the Japanese newt Cynops pyrrhogaster by carcinogen treatment. The carcinogen used was N-methyl-N'-nitro-N-nitrosoguanidine.

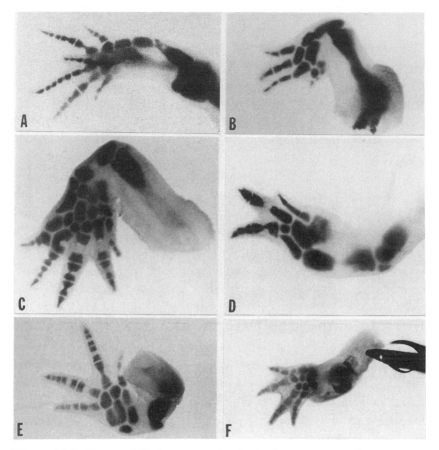

Figure 11.2 Abnormal limb regeneration in the Japanese newt *Cynops pyrrhogaster* induced by carcinogens. A, C, E: abnormalities induced by 4-nitroquinoline-1-oxide. B, D, F: Abnormalities induced by N-methyl-N'-nitro-N-nitrosoguanidine. The level of amputation was the elbow.

mice. Most spectacular is the case of the absence of ulna and radius which is also the effect when HoxA 11 and HoxD 11 are lacking in mice (Davis et al., 1995). Tumor induction was something that was never observed, at least in the regenerating limb. The intriguing similarity between the blastema and cancer cells may be why carcinogens are not able to induce cancer in amphibian limbs (Tsonis, 1983). It could be, for example, that any transformed cells are forced to differentiate and participate in regeneration when found in the blastema territory.

These instances of induction of limbs are in contrast with some of our beliefs on the initiation of regeneration and pattern formation. If an injury is sufficient to induce regeneration, then there should be an alternative route that ensures regeneration despite amputation. The formation of such limbs, however, leaves a big gap when models for pattern formation are considered. If amputation is not necessary, how does the stump acquire all the positional values of the circumference, provided by the amputation plane, that are obligatory for regeneration and pattern formation (see Part II)? One explanation to consider is that both the transplants and the injury bring into contact cells with different positional information, thus creating disparities that can result in intercalation and subsequent regeneration (Chapter 16). It is true that large wounds can produce supernumerary limbs (Bryant and Baca, 1978).

Postembryonic induction is another interesting scenario in the development of amphibian limbs. But the induction of limbs by other tissues could in fact bear great developmental significance. The limb territory, during development, coincides with that of the urogenital system (Nicholas, 1955). This is interesting because tissues from that system have one of the greatest limb-inducing capacities. In addition, the embryonic mesonephros has the ability to form cartilage (Lash, 1963), and ablation of mesonephros inhibits limb development (Geduspan and Solursh, 1992). The chondrogenic region of the mesonephros is restricted to the area of the limb primordia. This portion of the mesonephros not only forms cartilage but contributes cells to limb cartilage. It is possible that these tissues secrete factors important for limb development, and this could explain the observed induction. If that is the case, then similar studies with cell lines from these tissues or with fractionated proteins could lead to the identification of such factors.

12

Stimulation and Inhibition of Regeneration

12.1 Stimulation

Information generated from the field of amphibian limb regeneration will eventually help us understand why higher vertebrates, including humans, are not capable of regeneration. This hope has been present since Spallanzani's initial writings. He ends his book with the following sentence:

> But if the abovementioned animals, either aquatic or amphibious, recover their legs, even when kept on dry ground, how comes it to pass that other land animals, at least such as are commonly accounted perfect, and are better known to us, are not endued with the same power? Is it to be hoped they may acquire them by some useful disposition? and should the flattering expectation of obtaining advantage for ourselves be considered as chimerical?

Such thoughts have provoked scientists to experiment on the stimulation of limb regeneration in animals that are not capable of such regeneration. We already have seen how an added nerve supply (Chapter 6) or current application (Chapter 10) can induce regeneration in frogs. Similarly, many researchers have believed that the induction of dedifferentiation is of paramount importance. Therefore, experimental histolysis has become one method of choice in induction of regeneration, as mentioned in previous chapters.

Polezhaev (1946, 1972) reported that he was able to prolong the regenerative ability of limbs of *Rana temporaria* by repeated amputations (see also Kollros, 1984). What he accomplished by this method is to keep a constant wound or trauma on the limb. Induction of trauma could induce some degree of dedifferentiation and, therefore, was taken as a sign of regeneration. Such trauma was achieved by Cherkasova by traumatizing the wound surface in *Rana* with deep pricks of a needle; this resulted in increase in the ^{3}H-thymidine labeling index and incorporation of ^{3}H-lysine (Cherkasova, 1974). Induction has been reported in anura and mouse limbs with NaCl treatment (Rose, 1942, 1945; Neufeld, 1980). Similar experiments were conducted by Michael and Aziz in 1975 using *Bufo regularis* reuss. Limbs of stages 57, 63, and 66 were amputated at the thigh level. Three days later, the apical skin was removed and trauma was induced. The result was distal outgrowth. Other investigators attempted to induce regeneration in frogs by treatment with growth factors such as FGF, but they had minimal success (Gospodarowicz and Mescher, 1981). FGF2, however, has been found to induce regeneration of the chick limb bud (Taylor et al., 1994). When stage 25 limb buds were amputated at the prospective distal zeugopod, regeneration of digits was observed after FGF2 was applied locally. Similarly, other scientists infused blastemal extracts or alkaline phosphatase into irradiated limb stumps of the newt. It was found that alkaline phosphatase infusion promoted the growth of conical structures that contained amorphous cartilage (Deck and Dent, 1970). Wallace (1981) discusses whether or not this type of stimulated regeneration is epimorphic. He maintains that there is insufficient evidence from the reported experiments to conclude that epimorphic regeneration is indeed stimulated. These experiments could now be repeated, because markers of epimorphic regeneration are becoming available.

12.2 Inhibition

Scientists usually believe that you can learn about a developmental phenomenon if a specific way can be found to inhibit it. If an agent with a known mode of action and target inhibits an event, the target can be consequently linked to the particular event. As in many different fields, these ideas have been applied in the amphibian limb regeneration system as well.

Initial experiments using X-irradiation concluded that such treatment of the limb destroyed its regenerative capacity permanently and the amputated limb regressed. This effect occurs indiscriminately of the age of the animal or the stage of limb development. In irradiated axolotl larvae there is no sign of differentiation. In addition, the X-rays induce dedifferentiation of the existing cartilage, leading to the virtual disappearance of the entire limb skeleton (Butler, 1933; Puckett, 1936; Brunst, 1950, 1961). Similarly, regression of the limb results after UV-irradiation. However, such effects can be recovered by illumination with visible light for a few days after UV-irradiation.

Extensive investigations have also been performed to determine the effects of several chemicals that are known to interfere with DNA synthesis, DNA replication and transcription, and protein synthesis (Wolsky, 1974).

12.2.1 Nucleotide synthesis and limb regeneration

Attempts to interfere with the synthesis of purines or pyrimidines can have severe effects on limb regeneration. Several chemicals have the ability to inhibit nucleotide synthesis. Of these, colchicine is the most widely used in several studies. Thornton (1943) noticed that high concentrations of colchicine completely inhibited regeneration. Such an effect was marked by excessive dedifferentiation, followed by regression and resorption of the stump. The regression was observed when the agent was administered immediately after amputation. Treatment at later times resulted in suppression of regeneration without regression of the stump. Such results were verified using tail regeneration as a system (Lehmann, 1961). Colchicine effects differ from the ones produced by X-irradiation, however, in that when colchicine is removed, the regenerative capacity is regained. It is conceivable that the inhibition of regeneration by colchicine resulted from the prevention of mitosis, possibly through binding to microtubules. The thick epidermal cap that is characteristic in many instances of inhibited regeneration was not present in such treatments with colchicine. Another chemical, vinblastine, with colchicine-like action, has different results on regeneration. Studies on tail regeneration have shown that this agent affects different tissues selectively, with most effect on the mesodermal tissues such as notochord and muscle (Francoeur, 1968).

Inhibition of nucleotide synthesis can be obtained by using folic acid antagonists, such as aminopterin. The addition of one carbon to the purine ring is mediated by folic acid acting as coenzyme. The effects of aminopterin have been studied in forelimb regeneration of the axolotl by Gebhardt and Faber (1966). Although growth retardation was evident, no significant inhibition was observed. However, the effects of aminopterin on limb morphogenesis were quite unique. The regenerates were formed with either more or fewer fingers, depending on the plane of amputation. Polydactyly occurred only when amputation was performed at the proximal level, whereas oligodactyly was the result of distal amputation. In fact, when aminopterin was given at 14 days post-amputation, the incidence of oligodactyly was 100 percent. Such an effect is possibly dependent on the stage of the blastema.

Other substances that can interfere with nucleotide synthesis are structural analogues, such as fluorouracil, which has been shown to have inhibitory effects on *Rana catesbiana* tail regeneration when given intraperitoneally in repeated doses (Wolsky, 1964).

12.2.2 Inhibition of DNA replication or transcription and limb regeneration

The principal agent used in such studies was actinomycin D, which selectively inhibits RNA synthesis from DNA templates and which also affects DNA synthesis. Actinomycin D binds to guanine and inhibits RNA polymerase by a steric effect. Higher concentrations are needed for the same effect on DNA polymerase. The effects of this substance have been studied primarily by Carlson (1967a, 1967b, 1969) in limb regeneration of the newt *Triturus viridescens* and the axolotl *Siredon mexicanum*. The results showed that actinomycin is inhibitory to regeneration, resulting in a small blastema and a thick apical epidermis. Since inhibition of regeneration was observed even when only the skin was treated, it seems that actinomycin D could have affected regeneration by interrupting a signal from the wound epithelium that is necessary for dedifferentiation or proliferation.

Some other experiments have provided evidence that exogenous RNA administration can stimulate regeneration in Mexican salamanders that received massive doses of irradiation and in amphibia that normally do not regenerate their extremities. These were primarily the results of Polezhaev (1959, 1966, reviewed in 1972). Smith and Crawford (1969) also observed that intraperitoneal injections of liver RNA could cause dedifferentiation and blastema formation in amputated limbs of the frog *Rana pipiens*. At that time, such work might have seemed bizarre and unexplainable. However, RNAs have been found to be present on cell surfaces and, therefore, they can affect cell communication. The mutant axolotl c develops with a heart that fails to form myofibrils and does not beat. Studies have shown that total RNA isolated from intact tissue can stimulate heart differentiation of mutant hearts *in vitro*. The rescued hearts show restored myofibril organization and contractions (Lemanski et al., 1993). The identity of such regeneration- or differentiation-inducing RNA remains obscure, and so does any mechanism involving effects of exogenous RNA.

Analogues of nucleotides, such as 6-mercaptopurine, or alkylating agents (which react with nucleic acids or enzymes) acting by replacing hydrogen with an alkyl group, have been found to inhibit tail regeneration in frogs and axolotls (Wolsky, 1974). However, their action in limb regeneration is poorly investigated and not understood.

12.2.3 Inhibition of protein synthesis and limb regeneration

Antibiotics are the compounds that exert the most effects on protein synthesis. Chloramphenicol was found to inhibit limb regeneration in *Triturus viridescens*. The regenerate remained at the blastema stage for at least 40 days. For some cases that proceeded further, abnormal regeneration was recorded (Burnet and Liversage, 1964). Puromycin had similar effects on regeneration. It caused severe retardation and resulted in the generation of small cone blastema. Some of the regenerated limbs showed abnormal morphogenesis (Liversage and Colley, 1965). Streptomycin administration was found to induce malformations such as oligodactyly. This antibiotic did not affect the formation of the epidermal cap (Procaccini and Doyle, 1972). An analogue of an amino acid, N-dichloroacetyl-DL-serine, was also found by Gomes (1964) to inhibit limb regeneration, but to a lesser degree. The analogue had little or no effect on the process of dedifferentiation and blastema formation.

12.2.4 Effects of carcinogens on limb regeneration

The ability of certain substances to induce cancer, coupled with the ability of salamanders to undergo dedifferentiation, have laid the groundwork for the study of the effects of these substances on limb regeneration. In fact, these studies might be the most complete and comprehensive as far as the number of different carcinogens and species is concerned (for review, see Tsonis, 1983). Carcinogens have been found to have adverse effects on limb regeneration. At high doses they result in inhibition of regeneration, and at very low doses they have no effects; at critical doses, however, they alter pattern formation considerably. The teratogenic effects of carcinogens affect mostly the structures distal to the amputation level. Common abnormalities include polydactyly, oligodactyly, missing ulna or radius, variations in number of carpals, or a combination of all of these. The initiation of regeneration is severely retarded, and such retardation is probably connected to the slow formation of the basement membrane (Tsonis and Eguchi, 1983). Interestingly, the different carcinogens used were not able to induce cancer formation within the regenerate. Such an effect was linked to the power that the animals possess for regeneration. Such a hypothesis is supported by the fact that when limbs of the frog were denervated, sarcoma induction was achieved (Outzen et al., 1976). The role of carcinogens, therefore, was studied in the intact limb where it was found that they could induce normal-appearing supernumerary limbs (see Chapter 11). Abnormal regeneration has also been reported using the teratogen thalidomide (Bazzoli et al., 1977).

13

Genetics and Limb Regeneration

The field of limb regeneration is awaiting a molecular breakthrough. Isolation of genes involved in this process is of vital importance for understanding the mechanisms of the phenomenon and its restriction to some urodele species. The identification and utilization of mutants is the most efficient way to isolate the gene(s) responsible for a certain trait. *Drosophila* genetics is the best example of how screening for mutations has led to the isolation of genes responsible for many developmental events. To apply such a concept in the limb regeneration field seems impossible, however. Salamanders are not cloned, and they do not mature and mate well in the laboratory; they do not, therefore, provide the best material for transgenic studies. Furthermore, someone will observe the effect of a mutation on limb regeneration only after amputation is performed. In other words, there is not a phenotype to help identify a mutant salamander. Such a task is at least not realistic.

The only species that is bred regularly in the laboratory is the Mexican axolotl *Amblystoma mexicanum*. Many different mutations have been cataloged in this species, but not one with an effect on regeneration. Attention, therefore, was drawn to mutants with affected limb development. Limb regeneration and development share similarities as far as patterning is concerned (Muneoka and Bryant, 1982). Such a property could render these mutants unable to regenerate. The most studied of such mutants is the axolotl mutant *Short toes*, first characterized by Humphrey (1967). The mutant develops with abnormalities in the limbs and urogenital system. The affected animals have cystic, disorganized kidneys with glomeruli and tubules distorted or absent. Similarly, the ducts are very small or defective, showing extensive disruption of the columnar mesothelial cells. The limbs are characterized by absence of phalanges (mild cases) to absence of digits, carpals, and ulna or radius (Washabaugh, Del Rio-Tsonis, and Tsonis, 1993; Tsonis, Del Rio-Tsonis, and Washabaugh, 1993). In fact, such manifestations resemble the ones shown in mice where the HoxA 11 and HoxD 11 genes had been affected by gene targeting (Davis et al., 1995). Humphrey (1967) had reported that young mutant animals (3 months old) show some degree of limb regeneration. This has also been reported in more extensive studies by Mescher (1993). More detailed studies, however, have shown that limb regeneration in older animals is completely impaired. Amputation of limbs in 6-month-old animals results in a small blastema, but differentiation and regeneration do not follow. The small blastema is characterized by the presence of extensive extracellular matrix. More dramatically, the basement membrane, built between the epithelial cells and the underlying mesenchymal blastema, is abnormal (Figure 13.1). The basement membrane is very thick and convoluted in the mutant limbs. At the same time, tail regeneration is normal in the mutant animal (Del Rio-Tsonis et al., 1992). This result indicates that impairment in the regeneration is not the result of a sick animal but is rather associated with the mutation.

The utilization of such a mutant could be of great benefit to the field of limb regeneration. It could provide the material for genomic studies in which abnormalities can be linked to genes with effects on limb regeneration. The several manifestations of the disease are also of interest in the field of limb development and regeneration. For example, association of kidney and limb tissues has been mentioned in several other studies (Lash, 1963; Geduspan and Solursh, 1992). The mouse *Limb deformity* gene is also expressed in the kidney, and the mutant mice have affected kidneys as well. Renal agenesis in this mutant is connected to deficient outgrowth of the ureteric bud (Maas et al., 1994). Other studies

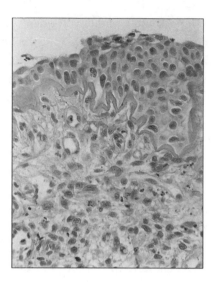

Figure 13.1 Limb regeneration is impaired in the mutant axolotl *Short toes*. This is a section through a blastema 3 months post-amputation. Only a hypomorphic cone has been formed without any sign of differentiation. The basement membrane is abnormally thick and convoluted.

have indicated that mesonephros is very important in inducing limb development. In experiments where the mesonephros has been ablated, limb development was not obtained (Geduspan and Solursh, 1992). This has been interpreted to mean that the progenitor cells that create all these different tissues are the same. Any mutation at that level of development can affect all these different tissues. True enough, the developmental territory for limbs and the urogenital system coincide in the embryo (Nicholas, 1955). Interestingly, kidney implants are one of the best for limb induction in amphibians (see Chapter 11).

Another mutant in which limb regeneration seems to be altered is the mutant axolotl *eyeless*. This mutant hypersecretes melanophore-stimulating hormone and hyposecretes gonadotropin. In addition, there is hypersecretion of prolactin-like hormones. All these neuroendocrine effects seem somehow to affect the rate of limb regeneration. In the *eyeless* mutant, limb regeneration is faster than in the wild type (Eagleson, 1993). Although these studies serve more to pinpoint the importance of hormonal balance in limb regeneration, the mutant offers a system whereby such effects can be studied.

PART II

PATTERNING IN THE REGENERATING LIMB

It is becoming increasingly clear that any consideration of the regenerative capacities of the urodele limb, especially in studies on the mechanisms of blastema formation and the establishment of a morphogenetic pattern, attention should be given to the fact that, although a blastema may be formed under highly atypical circumstances, it may undergo morphogenesis in a remarkably typical manner

—E.G. BUTLER, (1955)

14

Models for
Pattern Formation

The ability of the cells of the early embryo to differentiate into many different tissues and attain an identity in time and space is a fascinating aspect of embryology, and it may bear answers to many diseases as well. Not only do the embryonic cells differentiate, they also know their position as they create patterns unique to every tissue. How, for example, do the cells of the early limb bud build a hand with a certain number of bones and fingers, with a very precise shape? That the cells acquire a position tag is most notable in the regenerating limb where, after amputation, a faithful copy of the missing limb is always made. Scientists continue to be bewildered by such a regulative ability of the cells. The main question is how the pattern is regulated. How is the information provided for the discrete patterns to be laid down? What kind of rules govern such regulation?

Many different theories have been developed over the past decades to explain the mechanisms of pattern formation. The two major ideas are related to the positional information and prepattern aspects. Other ideas will be mentioned later. But first we will elaborate on theories that have used the regenerating limb as a model.

14.1 Positional information

According to this idea, formalized by Wolpert (1969), every cell possesses intrinsic information on its position in the body. The embryonic cells can acquire this information with a set of coordinates as they differentiate. The boundary coordinates are specified independently in the different parts of the embryo. The logical question is how the cells interpret the positional information. How, for example, are cells of the limb bud committed to cartilage, and which ones will go on to form an ulna and which ones will form a radius?

14.1.1 Gradients

Such specification can be provided by a chemical that can exist at a concentration that monotonically decreases (or increases) along a particular axis. Therefore, the cells can know their position along the axis by reading and interpreting the concentration. The gradient idea implies the existence of morphogens, chemicals, or molecules that exist endogenously in the body with the gradient properties, creating long-range diffusible signals. The evidence that retinoic acid might be a morphogen for limb development and regeneration will be presented in Chapter 17. The role of activin, however, as a morphogen during embryonic development has been shown as well (Gurdon et al., 1994). It was demonstrated that activin can be spread over a distance of 10 cell diameters by passive diffusion.

14.1.2 The boundary model

Proposed by Meinhard (1983), this model states that generation of positional information can be achieved at the boundary between two tissues by a diffusible morphogen that is produced at the boundary. Two boundaries exist for limb development, the anterior–posterior (AP) and the dorsal–ventral (DV). Cells at the distal tip of the AP/DV intersection (apical ecto-dermal ridge) produce a morphogen that sets the proximal–distal axis.

14.1.3 Polar coordinate model (PCM)

According to this model, championed by Bryant, Bryant, and French (1977), and Bryant, French, and Bryant (1981), the cells can assess their position from the coordinates of their immediate neighbors. This process does not require diffusible signals; rather, it relies on local interactions of cells. A simple representation of this model is illustrated in Figure 14.1. Each cell has information with respect to its position on a radius (A–E) and on a circumference (0–12). For example, in a limb, axis A represents the most proximal cells, and E the most distal; 12 represents the dorsal, 3 the anterior, 6 the ventral, and 9 the posterior value (see Figure 14.1).

Under this model, three rules should be observed. First is the complete circle rule, which states that a complete set of circumferential val-

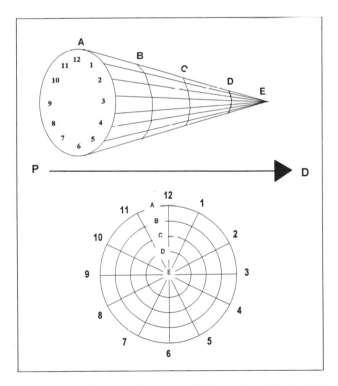

Figure 14.1 Polar coordinates of positional information. Each cell is assumed to possess information with respect to its position on a radius (A–E) and on a circumference (0–12). If this represents a limb, for example, A will be the most proximal and E the tip of the fingers. (Adapted from Bryant et al., 1981.)

ues should be present. This rule was modified (Bryant et al., 1981) to explain experiments that violated it (Chapter 16). The second is the shortest intercalation rule, that is, when cells from normally nonadjacent radial or circumference values are meeting each other (by amputation or grafting), the discontinuity stimulates cell division, and the positional values that are generated are the ones that lie between the two confronted values, using the shortest route. The third rule is the rule of distal transformation, which prohibits generation of more proximal radial values. This model explains several aspects of morphogenesis during limb regeneration that will be discussed later. One major point of this model is that it also links pattern formation and growth.

14.1.4 Progress zone model

This model, mainly supported by experiments with the developing chick wing, proposes that positional values along the proximal–distal axis depend on how much time the cells spend in the growing tip of the bud. Cells that spend the least time in the progress zone become proximal structures, while cells that spend the most time become distal structures. This model differs from the others in that positional information is dependent on growth and time (Summerbell, Lewis, and Wolpert, 1973). The model accounts for limb regeneration as well. Accordingly, the rapidly proliferating progress zone lays down the skeletal elements. The various skeletal elements are the same size as they leave the progress zone. The final differences in size reflect differences in the final growth phase. According to this idea, the total time for regeneration should be constant from different levels of amputation (Smith et al., 1974).

14.1.5 Serial threshold theory

Proposed by Slack (1980), this theory is similar to the PCM. Here, the pattern is generated by switching on and off genetically controlled regulatory substances within discrete mosaic regions in the limb.

14.1.6 The averaging model

According to this model, proposed by Maden (1977b), positional values can be averaged and by doing so the intercalary regeneration can be completed. Positional values from 0 to 100 are given to the limb along the proximal–distal axis. Accordingly, the wound epithelium would have value 0. If, for example, the limb is cut at position 60, there will be intercalary regeneration between value 0 and 60. However, the value for the dedifferentiated tissue will be 61. The model allows several steps for dedifferentiation of the more proximal structures whose values take place in the averaging. The confrontation of these values will initiate intercalation and dedifferentiation, then the process stops and differentiation and proliferation begin.

14.2 Prepatterns and deterministic mechanisms

The foundations of ideas about prepatterns were established from experiments with lepidoptera wings. In these structures, some color patterns are preceded by prepatterns in the rate of wing scale maturation. But how are the prepatterns set?

14.2.1 Reaction–diffusion models

In 1951, Belousov undertook the task of creating an inorganic equivalent of the citric acid cycle by oxidizing the citric acid not with the catalytic enzymes, but with the metal that these enzymes carry in their active sites. He selected cerium (because it gives a yellowish color) and inorganic bromate in a solution of sulfuric acid as the oxidizing enzyme cofactor NAD. The result was the first chemical oscillator. The yellow color from the oxidized cerium ions faded while they oxidized the citric acid, and then the color returned, as bromate ions oxidized the reduced cerium. This was literally the generation of a periodic pattern in a homogeneous chemical system (Winfree, 1987).

Independently, at about the same time, Turing (1952) was developing the chemical basis of morphogenesis by suggesting that an initially homogeneous system can develop waves in appropriate reactions involving autocatalytic and cross-catalytic interactions of two chemicals diffusing within the system. Turing proposed that these reaction–diffusion models could account for the generation of complex structures during morphogenesis and that these models in fact could be the setting conditions for the prepatterns. Periodic waves have been seen during fertilization (Ca^{2+}) (Lechleiter et al., 1991), and during aggregation of the slime mold (cAMP) (Martiel and Goldbeter, 1985). Reaction–diffusion can in fact be simulated by employing cell–cell communication via contact. These are the so-called cellular automata where the state of a cell can be determined in time t by the state of the neighbor cell in time $t - 1$ and according to a specific rule. Cellular automata can generate fractals, structures with similar patterns at different scales – something found in blood vessels or in lungs.

14.3 The topobiology principle

Developed by Edelman (1988), this theory claims that among neighboring and interacting cells there is influence because of the interaction. Cells scan and communicate with their environment via CAMs (cell-adhesion molecules) and SAMs (substrate-adhesion molecules). Depending on the location, cells can change CAMs or SAMs and this can specify position in the developing embryo.

In my opinion, any one model alone is not able to offer a solution to the complex and diverse structures in the body. It could be possible that the different models can be applied at different developmental times. For example, an initial wave mechanism can cause the first fundamental prepattern in the fertilized egg. This wave can then produce other waves through reaction–diffusion. These patterns can be periodic, but they can also develop to quasiperiodic or even chaotic by an external periodic force (Tsonis, Elsner, and Tsonis, 1989). The system can then start forming boundaries, and positional information can be acquired. Alternatively, cellular interactions within the different entities or among the different entities can lead to selective adhesion and rearrangements involving cell death and other types of selection. This could lead to distinct cell lineages. In this sense, cell-to-cell interactions involving ligands and receptors could supply positional information to the cells (Tsonis, 1987). As will be discussed in the following chapters, patterns in the regenerating limb could be well explained primarily by the polar coordinate model and by diffusible signals. Other ideas, however, could be valid as well.

15

Morphogenetic Properties of the Blastema

Upon amputation, the blastema can faithfully regenerate the missing structures distal to the amputation plane. The mechanisms whereby the undifferentiated blastema gives rise to the distal pattern have been the topic of research since the beginning of the century. More precisely, does the blastema depend on the parent differentiated tissues to assume its morphogenetic role, or is it an independent, self-organizing entity? Is the blastema pluripotent or not? These are the questions that I will try to address in this chapter.

15.1 Pluripotency versus self-organization

Scientists have tried to find answers to these questions, mainly by testing the morphogenetic potential of the blastema upon transplantation. The idea is that if the blastema is pluripotent, then it should exert appropriate properties when transplanted into neutral (that is, back of the animal) or foreign (that is, tail or eye) regeneration territories. If the blastema is pluripotent, it should differentiate according to the neighboring intact tissues, possibly by receiving signals from them that render it dependent. On the other hand, if the blastema is not pluripotent, such transplantations should result in generation of limb structures, and the blastema should be considered an independent and self-organizing system.

That the blastema cannot self-organize was initially suggested by *in vitro* experiments where the cultured blastema was shown to be unable to differentiate to limb structures (Schaxel, 1922). In other words, the blastema was dependent on parental tissues. In such experiments, however, the culture environment might have been limiting, thus rendering the blastema unable to progress further in regeneration.

Little is known about the behavior of limb cells in culture; only when this is well developed and controlled will such studies provide valid information. In fact, experiments that oppose the previous findings have been presented by Skowron and Roguski (1958). They observed that when dissociated cells were implanted into amputated *Siredon mexicanum* limbs that had been X-rayed, limb structures could be regenerated. Participation of limb blastema cell lines in newt limb regeneration has also been documented using modern techniques (see Chapters 4 and 8).

Some support for blastema pluripotency comes from transplantation studies into nonregenerative territories. Failure of blastemata to differentiate was reported by Schaxel (1934) and by Di Giorgi (1924) after they transplanted undifferentiated blastemata to the back of the animal. When the blastema was grafted with a portion of the stump or when stumps were grafted alone, however, normal regenerates with digits resulted. Detailed histological studies were not provided, and these experiments have been treated with skepticism. Resorption of the grafted tissues is often seen, and markers were not included to trace the fate of the blastema cells. Other experiments of this kind were performed by Hertwig (1925, 1927) and by Oka (1934). Hertwig transplanted the forelimb bud from a haploid Triton larva into the flank region of a diploid host. He was able to observe partial resorption, which somehow was followed by the production of limbs, which consisted of host tissue, as judged by the diploid cells, or of a mixture of both donor and host tissues. Oka transplanted limb buds near the limb field of different amphibian embryos and observed supernumerary limbs of host origin. The problem with such experiments, however, could be the induction of limbs in the embryo by the implantation procedure itself.

A somewhat stronger case for pluripotency and dependence is presented from experiments in which blastemata were transplanted into other regenerative territories. Exchange of forelimb and hindlimb blastemata resulted in limb regeneration according to the host. However, this dependency was related to the stage of the blastema. A critical stage, about 18 days after amputation, determined the fate of the regenerate. Beyond that stage, regenerates developed according to the donor blastema as well. This indicated that during the critical stage there was a response to induction by the host, but at later stages that was not the case. Such a dependence on the stage of the blastema was also observed in other experiments, which involved transplantation of tail blastema into the limb. Weiss (1925, 1927, 1930) reported that such transplantations could give rise to limbs, which led him to conclude that the tail blastema was induced to develop a limb by the host. This ability was gradually lost when older blastemata were transplanted. Intermediate blastemata gave rise to limbs and tails, whereas old blastemata produced only tails. The same criticisms about resorption can be applied to these experiments as well. In fact, the generation of a limb by tail blastema transplantation can be interpreted as (1) resorption of the transplant and (2) limb development by nerve deviation due to the injury. Schotte and Harland (1943) have shown that if a transplanted blastema does not fit perfectly, it is usually resorbed.

Perhaps even more intriguing experiments of this kind were reported by Farinella-Ferruzza (1950, 1953, 1956), who transplanted tail blastema of urodele embryos into the flank of anuran hosts. All the grafts formed tails. But, curiously enough, several weeks later limb development could be observed at the site of the grafting. These limbs were produced at the site from host tissues, or they could have been chimeras. More interesting, however, was the transformation of the induced tail into a limb after degeneration due to metamorphosis. If the tail resorbed, the cells of the flank could come together and form a limb, because supernumerary limbs can be formed on the flanks of amphibian embryos by wounding between limb fields. An alternative experiment involved transplantation of limb blastema into the tail. Guyenot and Schotte (1927) and Guyenot (1927) reported that when a 45-day-old blastema, at the early digit stage, was autografted to the tail, the digit buds regressed and the blastema slowly changed to a tail. Some other experiments involved transplantation of limb and tail blastema into lentectomized eyes (Schotte and Hummel, 1939; Trampusch, 1966). This work claimed that in more than 30 percent of the cases, transformation to lens occurred, whereas the control experiment exhibited less than 10 percent transformation. Again the weak point of these experiments was the lack of markers. When these experiments were repeated by Stone (1966), this transformation was correlated with the ability of amphibia to regenerate the lens and with incomplete removal of the lens, rather than with transformation of the limb and tail blastema to the lens. In fact, some urodeles would regenerate the lens from lens epithelium, and such remnants could stay behind after lentectomy. In another series of experiments, by Emerson (1940), there was no transformation in more than 300 cases involving tail blastema transplantation to the lens or ear region of embryos.

As already mentioned, all these experiments, though very stimulating, lack conclusive evidence because of the absence of a traceable marker to identify conclusively the origin of the regenerate. It is possible that a young, grafted blastema can be resorbed more easily than an old blastema. Therefore, as accounted in these experiments, transplantation using early blastemata gave results supporting the pluripotency aspects, whereas the same experiments with older blastema (which was already differentiating) supported the independence of the blastema. Such experiments could actually be repeated now with markers and some of these discrepancies can be solved.

But can a tail be transformed into a limb? The answer is yes. But this did not come from experiments involving transplantation of the blastema. The first transformation of a tail into a limb or gill-like structures was obtained in experiments conducted by Holtfreter (1955). He was able to obtain induction of ventral tail when a piece of kidney was implanted at the gastrula stage of a developing embryo. Amputation of such tails resulted in regeneration of limb or gill-like structures. This transformation is reminiscent of the one observed by Farinella-Ferruzza, who also achieved limb regeneration after amputation of tails induced by tail bud transplantation from *Triton* and *Amblystoma* to an anuran embryo at a neurula stage (1950, 1953, 1956). Such regeneration is known as heteromorphosis or homeotypic regeneration, now often called homeotic transformation. The earliest example reported could be the regeneration of a limb from an amputated antenna (Herbst, 1901). Perhaps most spectacular are cases of multiple limb generation at the site of tail amputation after treatment with vitamin A in premetamorphic frogs (Figure 15.1) (Mohanty-Hejmadi, Dutta, and Mahapatra, 1992; Maden, 1993). The effects of vitamin A (retinoic acid) on limb regeneration will be examined in detail in Chapter 17; nevertheless, such transformation is remarkable and might be related to the pluripotency of the regeneration blastema. The mechanism by which such a transformation occurs is unknown, but it definitely tells us that the morphogenetic field of a tail can be changed. But while these results support the idea that a blastema can be pluripotent and can change according to a stimulus, at the same time they contradict the idea of dependency from the host, which has been a prerequisite for pluripotency. It is possible that pluripotency and host dependence are two different things.

Support for partial or full self-organization for the blastema comes from a different group of papers. In 1932, David transplanted undifferentiated blastema with the stump from the upper limbs to the mid-body wall and observed formation of autopodia and some zeugopodia, but not stylopodia. Blastema from all stages without stump formed only autopodia, implying resorption of proximal tissues. Five years later, Polezhaev reported similar findings (1937, 1979). He transplanted undifferentiated

Figure 15.1 Homoeotic transformation of a tail into limbs in *Rana temporaria*. (Courtesy M. Maden.)

tail blastema to amputated limbs. When he transplanted one blastema, he observed limb formation in about 40 percent of cases and chimeric limb and tail fin structures in 60 percent of the cases. But when the transplantation involved more than four blastemata, tails or chimeras were seen in most of the cases. The opposite type of transplantation, that is, limb blastema to tail, showed similar results. More than one limb blastema was necessary to elicit formation of limbs. Such experiments imply that morphogenesis depends mostly on interactions among blastema cells and not with the stump. If parental interactions are taking place, there must be determination at earlier blastema stages.

The dependency of self-organization on the quantity of the transplant was also confirmed in experiments performed by Mettetal (1952). When he transplanted limb blastema with the stump, he found that resorption occurred in 16 out of 26 cases. In the remaining 10 cases limbs developed, but only in 2 were stylopodia formed, and zeugopodia were formed in 7. These results strongly indicate that stump and proximal blastema tissues were resorbed. When the transplantations took place without the stump, late blastemata gave rise to development of autopodia, but zeugopodia and stylopodia were developed only when the blastema was at the palette to two-digit stage. Mettetal concluded that organization of the blastema depends on both quantity of the blastema and stump inductive factors. To circumvent the apparent problem (that is, blastema independence of the stump), Faber (1960, 1962) suggested that the blastema has self-organizing and differentiation abilities for autopodia, but the development of zeugopodia and stylopodia needs signals from the stump. He derived such a conclusion from an experiment in which he placed carbon particles proximal to the apical epidermis. He then observed that the marker was found only in the zeugopodia and stylopodia. This experiment also suggested that the hand is derived from the most distal cells.

During the past three decades, however, a different approach has been applied to this kind of experiment. Small larvae, possessing a blastema that is rapidly grown and has genetic markers, are utilized to circumvent the problem of resorption. In experiments involving this kind of young blastema, the self-organizing properties of the blastema were shown more convincingly. De Both (1970) confirmed the aforementioned experiments by David, Mettetal, and Faber using undifferentiated axolotl blastema grafted to the flank. In these experiments, the mass of the blastemata was fused into a single regenerate, which in turn shows the importance of intrablastemal cellular interactions for morphogenesis to take place. In a more dramatic demonstration, Jordan (1960) observed that limb blastema from stage 52 *Xenopus* tadpoles was able to organize and form complete regenerates when implanted in the brain. Similarly, complete regenerates using *Xenopus* tadpole blastema were obtained after implantation in the eye chamber by Trampusch (1966) and by Dinsmore (1974). Stocum (1968a, 1968b) used blastema from *A. maculatum* larvae derived from the upper arm and transplanted it to the dorsal fin. Zeugopodia were developed in 70 percent of the cases. This incidence could be increased to 90 percent if the epidermis was stripped from the blastema before transplantation. These results showed that the blastema can self-organize and that this was under the influence of the epidermis.

Other experiments involved morphological, color, or genetic (ploidy) markers to demonstrate the self-organizing ability of the blastema. Stintson (1963, 1964a, 1964b, 1964c), using *Notophthalmus viridescens*, autografted the lower arm, or the anterior or posterior half of the lower arm, from a normal, unirradiated right forelimb into the upper arm of the irradiated left forelimb. As expected, the irradiated control never regenerated, but limbs that received unirradiated transplants were able to regenerate. The regenerated structures were not influenced by the host. They were developed according to the graft. Stintson obtained complete right forelimbs when the transplant was the whole lower arm, or half regenerates with graft asymmetry with the half anterior or half posterior lower arms. In similar experiments, *A. maculatum* blastemata from the upper arm were implanted into the tarsal level of the hindlimb. Nearly all regenerates developed zeugopodia, 50 percent contained stylopodia as well, and all were of forelimb origin. When triploid axolotl blastema from the wrist was implanted on the upper arm or thigh of diploid animals, the triploid cells were found only in the hand. The intermediate structures were filled by intercalation from the host stump. In other experiments, upper arm double anterior or double posterior blastema was exchanged with blastema from the normal upper arm. In the case of double anterior blastema, inhibited or truncated regenerates were observed, and in the case of double posterior blastema, double posterior or half limbs were generated.

In another series of experiments, the proximal region of a medium or palette axolotl blastema was used for transplantation (Michael and Faber, 1961). In one case, the blastema was implanted into the flank muscles. Even though at that stage cartilage differentiation had begun, it was shown that outgrowth of the blastema and formation of zeugopodia were preceded by dedifferentiation of the original cartilage. If the implant was grafted with the distal–proximal axis reversed, only digit differentiation was obtained in the majority of the cases. Similar results were obtained by Stocum (1968b) in a different experiment, using *A. maculatum*, wherein proximal blastemata were implanted into the dorsal fin. However, when proximal halves of redifferentiating upper blastemata were grafted into the tarsal level, the grafts were completely dedifferentiated and formed a complete regenerate in almost 100 percent of the cases (Stocum and Melton, 1977). These last experiments demonstrate that when blastemata are manipulated in the absence of a stump and again undergo dedifferentiation, they will self-organize. Thus it can be concluded that, at any stage, the blastema does not need signals from its differentiated tissue neighbors to exhibit its morphogenetic properties. Rather, these properties seem to be inherited by the blastema from its parent limb (Stocum, 1984).

An important point derived from this work (which spanned decades) is that the age of the blastema plays a significant role. For example, it appears that when early blastemata are used for the transplantations, they are not able to form complete regenerates; however older blastemata are able to organize. It could be that resorption is detrimental to morphogenesis of early blastemata but not to later ones.

15.2 Axis determination

The next question is whether axial polarity is affected by signals from the stump. Again, experiments involving transplantations were undertaken to answer this question.

15.2.1 The anterior–posterior (AP) and the dorsal–ventral (DV) axes

Initial experiments by Milojevic (1924) and by Milojevic and Grbic (1925) involved contralateral or ipsilateral grafting of *T. cristatus* hindlimb blastema after the AP or DV axis was reversed. Young blastemata (less than 12 days) gave rise to regenerates with the polarity of the stump, whereas older blastemata (more than 12 days) produced regenerates of their own polarity. Schwidefsky later (1934) indicated that the axial polarity is inherited at about 16 days after removal of the limb and that after determination, grafting always produced supernumerary limbs bearing the handedness of the stump. Iten and Bryant (1975) presented a series of experiments in which they transplanted distal forelimb blastemata of *N. viridescens* to a more proximal level, and vice versa, after AP reversal. Their results indicate that polarity in the early blastema is not determined and that the axis can be of stump origin. At later stages, the AP axis of the blastema is retained. When a proximal blastema was grafted in a more distal level with AP reversed, the regenerate had serially duplicated elements. The opposite grafting (distal to proximal) resulted in no missing structures, indicating a role of the stump in the intercalation process.

In a series of experiments on the axolotl conducted by Stocum (1982), pigmentation was used as a marker to identify the donor from the host. When blastemata were grafted to the contralateral limbs with the AP or DV axis reversed, they all developed to regenerates with the handedness of their origin. Supernumerary limb production was also noticed. The same results were obtained using early and medium bud blastemata. Stocum's experiments suggest that the transverse axis is inherited by the blastema from the parent limb (Stocum, 1984).

15.2.2 The proximal–distal axis

In this type of experiment, limbs were amputated distally or proximally and implanted in a pocket prepared in the body wall or the tail. Once the wound healed and the implanted limbs were firmly in place (vascularized and innervated), amputation was performed. Therefore, the limb had a free surface, but one with the opposite PD direction. Regeneration of these reversed limbs resulted in structures normally found only distal to the original amputation of the unreversed limb (Butler, 1955; Deck and Riley, 1958). In a different set of experiments, the distal part of a late bud blastema from *Amblystoma mexicanum* was removed and the remaining part was reversed. By doing so, a distal zeugopodium level was juxtaposed with a mid-stylopodium, thus creating a discontinuous PD axis. Such transplantations resulted in regenerates with the original polarity in more than 70 percent of the autografted (same limb, same level) cases (Michael and Faber, 1961; Faber, 1962). Oberheim and Luther (1958) inserted the hindlimb of *Salamandra salamandra* into the body cavity under the skin in such a way that the tip of the limb up to the wrist came out of the pocket. The limb was then amputated at the wrist level (not reversed) and also at the proximal humerus, where the polarity was reversed. Both amputations gave rise to distal elements that would normally regenerate from the respective amputation level. These results are similar to the transverse axis and indicate that blastema cells do not have fixed polarity.

16

The Regeneration of
Positional Information

The faithful regeneration of distal structures upon amputation implies that the cells along the limb know their position, or rather that they acquire the knowledge of their position upon amputation. Since only structures distal to the amputation plane can be regenerated normally, this means that the amputation plane determines the proximal boundary. On the other hand, evidence indicates that the wound epithelium imposes the distal boundary. The role of the wound epithelium and its importance for limb regeneration have been discussed previously (Chapter 3); however, the action of the wound epithelium as a distal boundary has been shown in the experiments discussed in this chapter. When denervated and epidermis-free blastemata from the upper arm are implanted completely into the dorsal fin, the regenerates are small and have stylopodia and zeugopodia, but not complete autopodia. But when the blastemata are implanted in such a way that their distal tips are exposed and covered by wound epithelium, the regenerates are small but complete in their distal structures. In the first experiment, their correct intercalation of positional values did not occur because the most distal part (wound epithelium) was absent; in the second experiment, correct distal structures were formed because wound epithelium was allowed to be reestablished (Stocum and Dearlove, 1972).

As mentioned in Chapter 14, the regeneration of positional information during limb regeneration can be explained within the context of the polar coordinate model (PCM). Also applicable are concepts involving diffusible morphogens. As explained earlier, the PCM assumes that local cellular interactions suffice for the acquisition of positional information. This view is in contrast with the role of morphogens, which can act through a long range. The PCM model has its roots in pioneer experiments conducted by Harrison (1921) and by Nicholas (1924a, 1924b) (see also Korschelt, 1931) that dealt with the effects of transplantation of limb buds after rotation of the anterior–posterior or dorsal–ventral axis. These investigators had observed that when such transplantations were performed, supernumerary limb formation resulted. Even though such supernumerary structures could be explained by the rules of Bateson (1894) (which instruct that the supernumerary limbs are usually mirror images of the original), these phenomena were not fully understood until the concepts of the polar coordinate model were laid down. Let us now consider the evidence that contributes to the major ideas about the regeneration of positional information.

16.1 The rule of shortest intercalation

According to this rule, when two cells of different radial or angular values come into contact, either by grafting or wound healing, the discontinuity in positional values stimulates cell proliferation and growth and leads to the generation of positional values that normally lie between the values of the confronted cells. For example, if B and D values are confronted, intercalary regeneration will result in the generation of all values between B and D. Similarly, if position 1A is confronted by 4A, intercalation will result in the generation of 2B and 3B in the first step of growth. In other words, confrontation of angular values will result in the shortest intercalation (2B, 3B) and not the alternative longer route

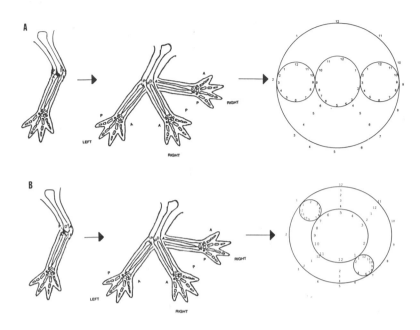

Figure 16.1 Illustrations depicting the contralateral (A) and ipsilateral (B) transplantations of blastema. In the first series, a blastema from a left limb is grafted onto the stump of a right limb. This procedure reverses the anterior–posterior (AP) axis. The result is generation of two supernumerary limbs that are mirror images of the graft. Such generation can be explained by the PCM. The outer circle indicates the stump and the inner circle the graft. Because the AP axes are opposed, the only possible intercalations that can give complete circumference (according to the rule of shortest intercalation) are at the positions of the two circles. Other confrontations will not produce a complete circumference. In the second series, the blastema is rotated 180° and regrafted on the same limb. This ipsilateral transplantation will reverse the AP axis as well as the DV axis. The result will be two supernumerary limbs at predictable positions. This can also be explained by the rules of the PCM. The only position at which complete circumference can be generated will be at the posterior dorsal quadrant (value 1 of the stump confronting value 7 of the graft), or at the ventral anterior quadrant (value 7 of the stump confronting value 1 of the graft). (Adapted from Bryant et al., 1977.)

(5B, 6B, 7B, 8B, 9B, 10B, 11B, 12B) (see Figures 14.1 and 16.1). When a distal blastema is implanted in the proximal host, regeneration proceeds distally with the donor blastema forming the distal structures of its origin and the host filling in the missing values by intercalation. Since the wound epithelium has value 0, amputation of the limb along any point will lead to regeneration of the missing distal structures by this intercalation rule. The rule of distal transformation, which restricts regeneration of more proximal structures beyond the amputation plane (see Section 16.3), is not violated by intercalary regeneration. The requirement of shortest intercalation imposes computation of the distance between positional values. It is not known how cells do such measuring. In addition, limbs with no apparent continuity can be regenerated (see Section 16.2.5). This phenomenon violates the rule, implicating other ways for the specification of positional information (Glass, 1977; Holder and Weekes, 1984).

16.2 The rule of complete circumference

According to this rule, all the angular values (circumference: 0–12) must be present for regeneration to take place. If any of these values is not present in an amputation plane (an experimental circumference), regeneration cannot take place. Lheureux (1975) showed such a requirement by transplanting skin in irradiated limbs. As we already know, an X-irradiated limb loses its ability to regenerate, but transplantation of skin can restore this ability (Trampusch, 1956). The skin, however, will not be able to restore the regenerative capacity unless it contains enough circumferential values to create positional disparities. When skin with one positional value, say only dorsal, was grafted, no regeneration resulted, but when values from dorsal and ventral regions were included, regeneration occurred. The PCM model came under severe criticism after it was found that constructed double anterior or double posterior limbs regenerated distally. Similarly, the development of axolotls with double posterior limbs and their subsequent successful regeneration argued against the validity of this rule. The revised version of the PCM, however, implicated the mode of wound healing as a necessary element for the regeneration of limbs without complete angular values by intercalation. According to this, depending on the wound closure that establishes cell-to-cell communication and interactions, a circle can be made by circumferential intercalation according to the shortest intercalation rule. Such a model can predict the regeneration of double anterior or double posterior blastema.

16.2.1 Regeneration of double posterior limbs

Initial experiments by Bryant in 1976 showed that double posterior or double anterior limbs failed to regenerate. The limbs in that case were amputated 3–4 weeks after construction. It now seems that, depending on the mode of wound healing, different outcomes can be obtained from the regeneration of double posterior limbs. Double posterior limbs were constructed and allowed to heal for different periods of time. Holder and Tank (1979) and Holder (1981) observed that whenthese limbs were amputated 10 or 15 days after construction, limbs failed to regenerate. However, when amputation was performed 5 days after construction, symmetrical double posterior limbs with posterior symmetrical elements were regenerated. In cases where the limb was subjected to a second amputation, more structures were regenerated (Figure 16.2). Regeneration of these limbs has been observed also by Stocum (1978, 1980, 1981) and by Slack and Savage (1978a, 1978b). It was reported that the limbs that were amputated early (10 to 14 days) were more likely to regenerate distally than the ones amputated later (32 days). Bryant et al. (1981) have proposed models of wound healing that could interpret these results. In the long-term healing, cellular interactions after construction of double posterior limbs allow the same positional values to come into contact; this would inhibit regeneration because of the lack of intercalation. When short-term healing is allowed, however, interactions between the same positional values are prevented and interaction of different positional values could result in intercalation and distal growth (Figure 16.2).

16.2.2 Regeneration of double anterior limbs

Regeneration of constructed double anterior limbs can also be obtained when the amputation happens soon enough after the construction to ensure proper cellular interactions. After amputation of such limbs constructed at the shank level, double anterior regenerates with symmetrical anterior elements were formed. Double anterior limbs amputated at the thigh were unable to regenerate, however (Stocum, 1980, 1981). This presents a nonequivalent potential for regeneration between thigh and shank double anterior limbs, which is in contrast with the PCM. Furthermore, in every experiment, the regenerated elements from the

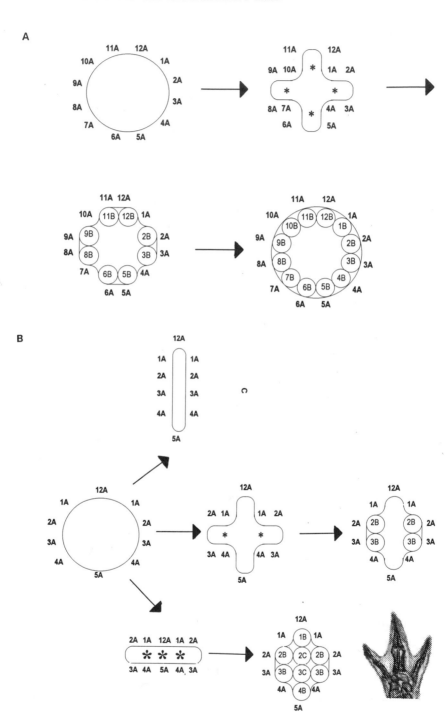

C

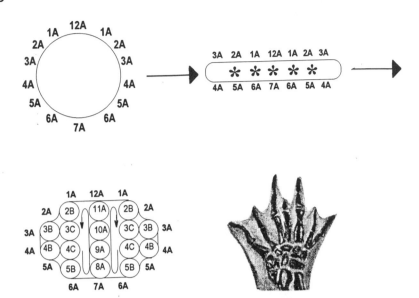

Figure 16.2 Mode of distal growth according to types of wound healing. A: In this experiment, removal of tissue has left an asymmetrical surface with radial value A. In this type of healing, the values represented with asterisks will be generated by intercalation. This will be the first step of distalization with some B values regenerated. Subsequent intercalation will produce all the values and will result in complete distal growth. B: This represents regeneration from a symmetrical surface (double half limb). Depending on wound healing (which is the way that the different values will come in contact), no growth or incomplete growth can be received. If, for example, it so happens that there is no confrontation, no growth will result because of the lack of intercalation (upper panel). But if the healing can follow a different mode, such as the one confronting some values, intercalation and distalization can result in some distally incomplete growth (middle and lower panel). A limb generated by such a mode is presented. Here we have a convergent regenerate formed from double posterior symmetrical stumps. C: A different healing mode can result in an expanded limb. (Adapted from Bryant et al., 1981.)

double posterior limbs were always more extensive than the ones in the regenerates from double anterior limbs. These differences can be explained if we assume that the spacing of the positional values is different in the thigh and the shank, and that most are carried in the posterior half of the thigh. The distribution should be unequal in the anteroposterior section of the shank as well, but more evenly distributed than in the thigh (Figure 16.3).

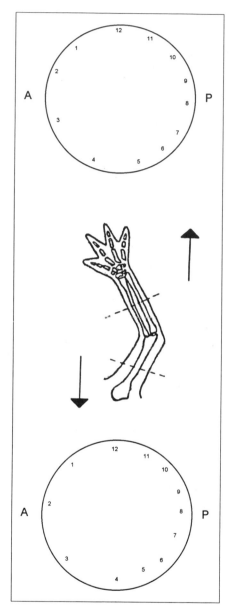

Figure 16.3 An illustration demonstrating the distribution of positional values in the anterior–posterior axis and their differences along the proximal–distal axis. The spacing of the circumferential values is different in the proximal region as compared to the distal. Most of them are carried in the posterior half of the thigh (proximal). They are also unequal in the shank (distal) but more evenly distributed than in the thigh.

16.2.3 Regeneration of double dorsal or double ventral limbs

Double dorsal limbs can regenerate distally in most of the cases. The regenerates consist of single limbs with DD muscle pattern or single limbs with mixed DV muscle pattern. Double ventral limbs also regenerate distally. All the regenerates consist of single limbs with ventral–ventral (VV) muscle pattern (Ludolph, Cameron, and Stocum, 1990).

16.2.4 Regeneration of half limbs

Half anterior or half posterior limbs can be constructed by removing half of the lower arm or leg through a mid-line between radius and ulna. Half anterior limbs can regenerate posterior structures. The degree of regeneration varies from complete regenerates to the regeneration of only some toes. Similarly, the regeneration of half posterior limbs proceeded to various degrees of completion. When half limbs are amputated, regeneration of the amputated half is restored. Likewise, half dorsal or half ventral limbs can regenerate the opposite structure to varying degrees. In the majority of the cases, however, each half failed to regenerate a complete complementary half muscle pattern (Bryant, 1976; Kim and Stocum, 1986; Ludolph et al., 1990; Stocum, 1991). When half upper arms were amputated immediately, they developed single complete regenerates. But if the amputation was delayed for more than 3 weeks, which permitted a complete wound healing, a supernumerary limb from the wound surface was developed in addition to the complete regenerate from the amputation surface (Bryant and Baca, 1978).

16.2.5 Rotation of grafted blastema along the AP and DV axes

Interactions such as the ones predicted by the PCM along the different axes were studied by transplanting blastemata to contralateral or ipsilateral limb after reversing their anterior–posterior (AP) or dorsal–ventral (DV) axes.

When a blastema is contralaterally transplanted with AP axis reversed, the DV axis stays the same. This type of transplantation has been performed with mid-bud blastema or with older blastema of the palette stage (Iten and Bryant, 1975). In a different series, the blastema was grafted from a distal to a proximal location or vice versa. The result from such transplantation is the production of supernumerary limbs. Because the AP axis is reversed, different positional values in the two halves are confronted. Intercalation from this confrontation leads to the development of supernumerary limbs arising from the posterior, anterior, or from both sections of the blastema–stump plane

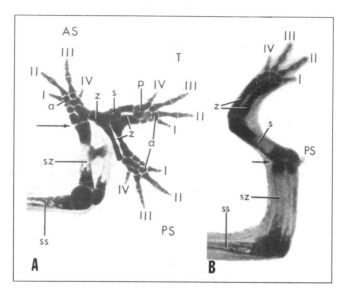

Figure 16.4 A: Supernumerary limbs resulted from transplantation of left proximal regenerate at the early digit stage to right distal stump. The autopodium developing from the transplant retained its left-handedness. Both anterior and posterior regenerates are right limbs. All three regenerates possess zeugopodial elements that articulate with the duplicated humeral head. Arrow indicates the level of transplantation. B: Skeletal preparation of a limb showing serial duplications of structures in the proximal–distal axis. A left proximal blastema at the late bud was transplanted to a right distal stump. PS: a small posterior supernumerary regenerate consisting of a single digit. Arrow indicates the grafting site. z: regenerated zeugopodium, s: regenerated head of the stylopodium, a: single anterior basal carpal of digits I and II, p: single large proximal posterior carpal, ss: stump stylopodium, sz: stump zeugopodium. PS: posterior supernumerary regenerate, AS: anterior supernumerary regenerate, T: transplant. (Courtesy S. V. Bryant.)

(Figures 16.1 and 16.4). Such results were obtained in the 1920s with transplantation and rotation of embryonic limb buds (Harrison, 1921; Nicholas, 1924a, 1924b). Similar transplantations using grafts from stage 34 suggested the existence of a zone of polarizing activity (ZPA) in amphibian limbs (Slack, 1976). One result from all different types and series of transplantation is that the frequency of supernumerary limb production is higher when older blastemata were used. The handedness of these supernumerary limbs was mostly of the stump when transplantation to the same level along the proximal–distal axis was performed (proximal to proximal, or distal to distal). When distal-to-proximal or distal grafting occurred, the handedness was of the stump, but there were also many intermediate cases. In all series, however, no supernumerary limb showed the handedness of the graft. The handedness of the limb, on the other hand, was that of the graft or intermediate when a late blastema was used, but mostly of the stump when a mid-bud blastema was used (Tank, 1978).

These experiments clearly showed that the generation of supernumerary limbs is dependent on the age of the grafted blastema and on the interactions and influence of the stump. The generation of such supernumerary limbs is predicted well by the PCM. When young blastemata were transplanted from proximal to distal with AP axis reversed, limbs were generated with serially duplicated segments along the proximal–distal axis. The opposite grafting, distal to proximal and AP reversal, did not result in deleted structures, and this did not correlate with the stage of the transplanted blastema (Iten and Bryant, 1975). Such data suggest that despite the ability of the blastema to self-organize, it does interact with the stump to produce the intercalary regenerate. When the dorsal–ventral axes are opposed in contralateral transplantations, without shifting along the proximal–distal axis, supernumerary limbs are produced with the same handedness of the limb stump. These limbs are mirror images of the regenerate that develops directly from the transplanted blastema (Bryant and Iten, 1976).

This is not the case, however, when ipsilateral transplantation was performed. Such an operation with a reversed AP axis produces a reversed DV axis as well. Supernumerary limb formation is also the result of such an experiment, for factors similar to those occurring from contralateral transplantation. These limbs were developed from the anterior and posterior sections or from the anterior section only. The ones that sprouted from the anterior had the handedness of the stump, but many were of mixed handedness as well. The ones arising from the posterior were opposite to the stump or mostly mixed. Apposition of the dorsal and ventral axes also led to the formation of supernumerary limbs of stump handedness sprouting from dorsal and ventral positions (Figure 16.5) (Maden, 1980; Thoms and Fallon, 1980; Tank, 1981; Maden and Mustafa, 1982). The PCM also predicted that, apart from 180° rotations, supernumerary limbs should not be obtained. This was not the case, however, when 90° and 270° rotations (Wallace and Watson, 1979) or

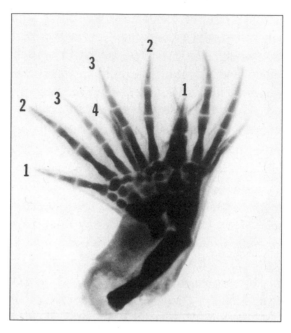

Figure 16.5 A double anterior supernumerary limb that is the result of two supernumeraries adjacent to each other and sharing a common fourth digit. These supernumeraries were induced by 180° ipsilateral blastema rotation. The digits of the supernumerary are marked. The rotated graft is on the right without digit numbers. (Courtesy M. Maden.)

45° and 315° rotations (Maden and Turner, 1978) were performed. The highlight of these experiments was that some of the resulting supernumerary limbs showed anatomical discontinuities: they were double dorsal or double ventral. Such a situation is in contrast with the PCM, which does not predict the possibility of such structures. This occurrence may indicate a mosaic behavior of the regenerating limb, with the different halves of the limb being organized independently. The problem of discontinuity was addressed by Muneoka, Holler-Dinsmore, and Bryant (1986a), using the fact that circumferential intercalation is directionally biased.

This bias can also be seen in proximal–distal transplantations of blastema. When a distal blastema is transplanted to a proximal level, intercalation occurs and a complete limb is produced (Pescitelli and Stocum, 1980). When a proximal blastema is transplanted to a distal level, however, graft and host fail to interact and produce a regenerate

(Iten and Bryant, 1975). Also, it has been found that when supernumerary limbs are produced by contralateral grafts, posterior cells contribute to posterior and dorsal regions, and anterior cells contribute to anterior and ventral regions of the supernumerary limbs (Muneoka and Bryant, 1984a, 1984b). Hence, there is a directional bias from posterior to dorsal and from anterior to ventral. Muneoka, Holler-Dinsmore, and Bryant (1986a) constructed limbs with two types of discontinuities. In type I, anterior–ventral tissue was opposed to posterior–dorsal (directional bias); while in type II, anterior–dorsal tissue was opposed to posterior–ventral. Regeneration of these limbs resolved the discontinuities in type I limbs only. These results indicated that the discrepancies arising from the anatomical discontinuities can be explained by the directional bias of circumferential intercalation.

A modified theoretical model, the hierarchical polar coordinate model (HPCM), has been proposed to circumvent these problems (Papageorgiou, 1984; Costaridis et al., 1989). According to this model, single circles, such as the ones used to predict the outcome of transplantation with the PCM, are not sufficient to reconcile the direction of intercalation being in coincidence (congruence) with the stump and the graft. But if twisted circles can be used, the continuity congruence is formed (Figure 16.6). This model can explain and predict the occurrence of double dorsal (Figure 16.7A), double ventral (Figure 16.7B), double posterior (Figure 16.8), or mixed limbs (Figure 16.9). The twisted circle allows the anatomical discontinuities without breaking its continuity (Figure 16.6). This model predicted that even in contralateral transplantations, someone could obtain double dorsal or double ventral limbs if large numbers of limbs were tested, because statistically the possibility of generation of these symmetrical limbs is very low. True enough and as predicted by the HPCM, these limbs were observed after contralateral transplantations (Figures 16.7–16.9) (Costaridis et al., 1989, 1991).

Supernumerary limbs can also be obtained after a 180° rotation of the skin or the muscle only. Such an experiment suggests that positional dislocation along the anterior–posterior axis rather than rotation of the axis leads to the formation of the supernumerary limbs. Rotation of bone or stump epidermis, on the other hand, did not result in supernumerary regeneration (Carlson, 1974b, 1975a). This is interesting because bone transplantation can rescue regeneration in X-irradiated

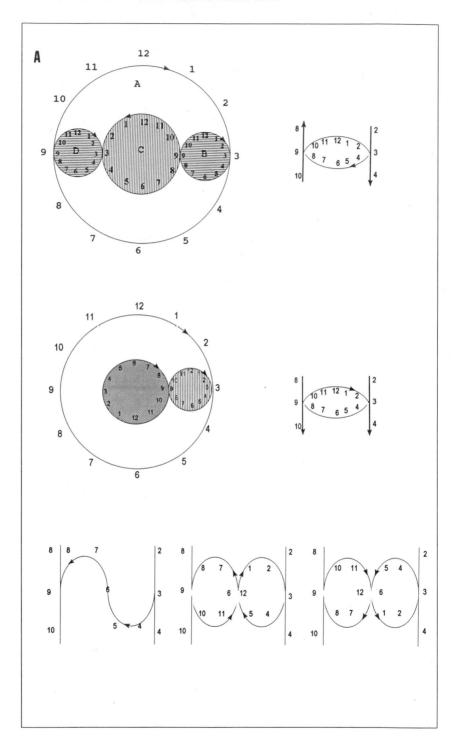

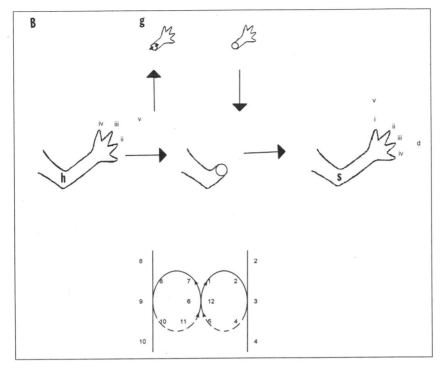

Figure 16.6 A: A schematic presentation of a contralateral transplantation indicating congruence. In this kind of experiment the intercalating paths of supernumerary limb B are congruent with the stump and the graft (upper). (A): Left-hand stump, (C): distal right-hand graft in dorsal and ventral coincidence with the stump, (B), D: two possible supernumerary limbs with the stump handedness. This can be better realized with supernumerary (B) where cell 3 of the stump confronts cell 9 of the graft. The intercalating contour is in coincidence with the one of the stump and graft (upper right). In ipsilateral transplantation, however, the intercalating paths of the supernumerary limb are congruent with the stump but not with the graft (middle panel). Such an intercalation path, however, can become congruent to both stump and graft if the twisted contour is used instead of a simple contour (lower panel). Such a twisted contour representation of the regenerated limbs can explain regeneration of double dorsal, double ventral, or mixed limbs observed especially upon ipsilateral transplantations. Such examples are presented in Figures 16.7–16.9. **B**: h is the host (dorsal side) and g is the graft (ventral side). Both AP and DV axis are confronted. s is the product, a supernumerary limb of mixed handedness (ventral side of digits i, and ii, and dorsal side of digits iii and iv). The twisted contour represents the HPCM explanation for the generation of such limb. (Adapted from Papageorgiou, 1984.)

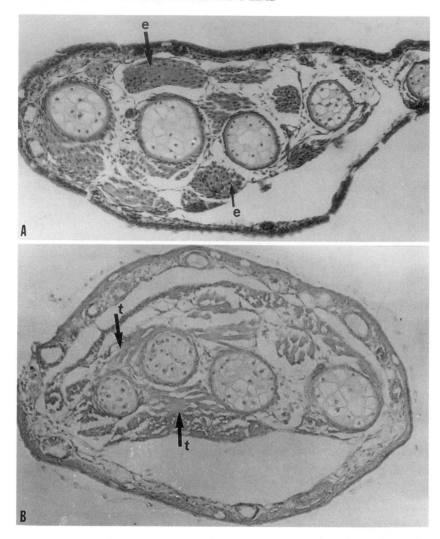

Figure 16.7 A: Double dorsal supernumerary limb generated by 180° rotation and contralateral graft of blastema in *Bufo bufo* larvae. The characteristic muscle (extensores breves, e) of the dorsal part appears as clusters and can be seen in the ventral part as well. B: Double ventral supernumerary limb from a similar experiment. The characteristic muscle (transversi metatarsi, t) of the ventral part can be seen in the dorsal part as well. (Courtesy S. Papageogiou.)

limbs (Trampusch, 1956). This might mean that the inductive mechanisms in regeneration are not linked with the patterning ones. A different study evaluated the ability of bones (such as humerus, radius, and ulna) to induce supernumerary limb structures. It was found that

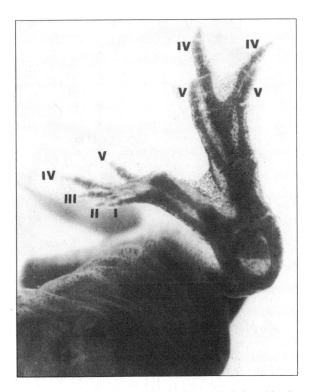

Figure 16.8 Dorsal–ventral reversal by grafting a left-hand bud on a right-hand stump in *Bufo bufo* larvae. The graft developed into a normal limb (left) while a double posterior structure was generated as well (right). (Courtesy S. Papageorgiou.)

humerus or radius can produce supernumerary digits only when grafted to either a posterior or dorsal site. In contrast, ulna grafts are capable of inducing supernumerary digits in all four host sites (Gardiner and Bryant, 1989). These results differ from those of Carlson (1975a), but this could be attributed to the fact that adherent connective tissues were included in the Gardiner and Bryant study. In other words, only these fibroblasts could carry positional information that would lead to confrontation and intercalation.

As evidenced from these experiments, the PCM can explain the generation of limb structures after transplantation and it is, therefore, useful in explaining the mechanisms of patterning during regeneration. This model can also account for similar data on limb regeneration in arthropods (Bryant et al., 1977, 1981). The model is very simple in its

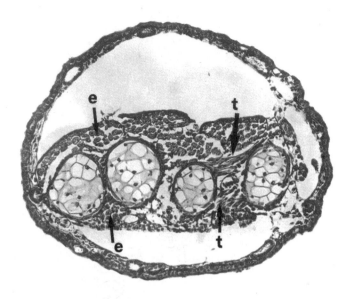

Figure 16.9 Transverse section of a mixed symmetric supernumerary limb at the metatarsal level, generated by contralateral transplantation inverting the dorsal ventral axis in *Bufo bufo* larvae. Note the characteristic muscle of the dorsal part (e) present in the ventral part and the muscle of the ventral part (t) on the dorsal part. The right part of this limb is double ventral and the left part is double dorsal. (Courtesy S. Papageorgiou.)

conception and by no means can it account for all observations, as already pointed out. In this sense, I would like the reader to keep in mind the concept quoted from Butler (1955) at the beginning of this part.

16.3 The rule of distal transformation

According to this rule, no element more proximal to the amputation level can be regenerated. Such an idea was proposed by Weismann (1892). Therefore, intercalary regeneration does not violate the rule of distal transformation. This concept has been verified by transplantation of distal blastema to a more proximal level (Pescitelli and Stocum, 1980). It has been observed, however, that vitamin A derivatives can, in fact, result in more proximal regenerates, thus violating this rule. Distal blastema treated with vitamin A can regenerate more proximal structures. This has enormous implications in the field of pattern formation and the underlying molecular biology of the field. A detailed account of the action of vitamin A will be presented in Chapter 17.

How stable is the positional identity in the cells that specify it? To answer such a question, Groell et al. (1993) cultured posterior blastema cells for several days and then transplanted them into the anterior region of a host blastema. It was found that these dissociated cells were able to induce the formation of supernumerary skeletal elements even after being in culture for a week. This is interesting because we now know that there is a timeframe during which manipulation and study of the positional identity can be studied in culture.

16.4 Evolution of the regeneration of positional information

Morgan (1901) identified two different types of regeneration. Epimorphosis describes the mode by which proliferation precedes the development of the new elements. In morphallaxis, however, a part is transformed into a new part without proliferation at the amputation surface. Morphallactic regeneration has been described in *Hydra* and *Dictyostelium*. Because of the direct transformation of the remaining part into a new one without proliferation, such a mode of regeneration results in a pattern with variable size. In contrast, epimorphosis results in patterns where the size is always the same, regardless of the amount of removed tissue. Studies of amphibian limb regeneration with different sized blastemata indicated that in every case the number of the regenerated elements is the same, but their size varied according to the size of the blastema (Maden, 1981b). This is, in fact, a morphallactic behavior in an epimorphic system. Maden hypothesized that the same mechanisms of pattern formation were preserved during the evolutionary transition from morphallaxis to epimorphosis. Growth did not replace the mechanisms of pattern formation, but was added to the existing ones. In this scenario, the regeneration of positional information during limb regeneration is governed by an early phase of morphallactic change in the positional values before the onset of growth.

17

Vitamin A and Patterning

The first indication that vitamin A compounds (such as retinol and all-*trans*-retinoic acids) can affect morphogenesis in the amphibian limbs came from the experiments of Niazi and Saxena (1978). These investigators found that treatment of frog tadpoles with retinol palmitate can result in abnormal limb regeneration. Maden (1982) presented a detailed analysis on the effects of vitamin A compounds on axolotl limb regeneration. The ability of these substances to produce duplications along the proximal–distal axis was verified. Usually amputation through the wrist results in regeneration of only the hand elements. The same amputation, however, will result in additional, more proximal regeneration after treatment with vitamin A (Figure 17.1). Maden noticed that the effect of vitamin A was also dose-dependent. The larger the dose of vitamin A, the more proximal elements could be regenerated. After a critical dose, however, the proximalizing effects of retinoic acid are replaced by inhibitory ones. The effects of vitamin A can be observed by placing the animals in a solution, by injecting them, or by vitamin-A–carrying bead implan-

154

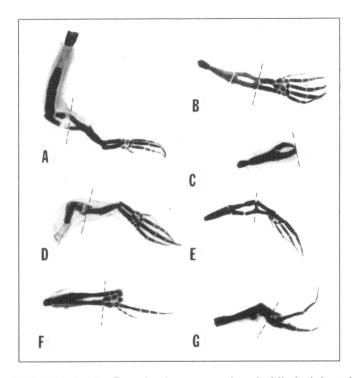

Figure 17.1 Victoria blue B–stained regenerated axolotl limbs injected with all-*trans*-retinoic acid, 100 µg/g body (A, G); 9-*cis*-retinoic acid, 100 µg/g body (B, C, D, F); or 9-*cis*-retinoic acid, 50 µg/g body (E). A, B, D, E: proximalization of the regenerates, C: inhibition of regeneration; F, G: abnormal regenerates. The dashed lines represent the level of amputation.

tation directly into the limb (Maden, Keeble, and Cox, 1985; Stocum, 1987; Stocum and Maden, 1990). The necessary doses in each type of treatment have now been well documented (Thomas and Stocum, 1984).

This property of retinoids provided the foundation for the school of thought that positional information could be provided by a chemical that exists as a gradient in the tissue. The cells along the axis of the tissue (in this case the limb) will know their identity or position by interpreting the concentration of the chemical. The dose-dependent effect of vitamin A suggested that the vitamin is present as a gradient along the proximal–distal axis, with more of it in the proximal tissues than in the distal. In other words, cells of the humerus will respond to more vitamin A than cells of the carpals. Therefore, when carpal cells are treated with vitamin A, they are exposed to more vitamin A levels and, therefore, they behave like humerus cells. Hence, proximalization results: a humerus, ulna, and radius are regenerated along with the parts of the hand. Such results have also been obtained with an analogue of all-*trans*-retinoic acid, the 9-*cis*-retinoic acid (Tsonis, Washabaugh, and Del Rio-Tsonis, 1994). In fact, 9-*cis*-retinoic acid is more potent in eliciting a morphogenetic effect, which could bear importance when molecular mechanisms are taken into account (see Section 17.2). Obviously, such an ability of retinoic acid violates the rule of distal transformation and provides the framework for gradients involved in pattern formation. The proximalizing effects of vitamin A have been best studied by immediate treatment of the amputated limb or by treatment after 3–4 days. Johnson and Scadding (1992) have shown, however, that amputation 2 weeks after treatment can result in proximalization. Concentration of retinoids in the body of the animal is still higher at the time of amputation, indicating persistence of injected retinoids in the body.

Retinoic acid can also proximalize the differential affinity of the blastema. When the blastema from a forelimb, wrist, elbow, or mid-upper arm is transplanted into host hindlimbs regenerating from the mid-thigh, the grafts are displaced to the corresponding host levels. This indicates differential affinity for the blastema from different levels. Treatment with retinoic acid proximalizes the affinity of distal blastema. These results provide evidence that positional memory in regenerating limbs is related to blastema cell affinity (Crawford and Stocum, 1988).

Nevertheless, the proximal–distal axis is not the only axis that retinoic acid can affect. In experiments designed specifically to address this question, half limbs were constructed. When half anterior limbs were treated with retinoic acid, posteriorization of regeneration resulted with the posterior elements present, while similar treatment of half posterior limbs resulted in inhibition of regeneration. This means that retinoic acid could modify positional memory in one direction along the anterior–posterior axis, namely, the posterior (Kim and Stocum, 1986). Similar experiments with half dorsal and half ventral limbs indicated that retinoic acid ventralized the positional memory along the dorsal–ventral axis. Half dorsal treated limbs restored the dorsal–ventral axis, while half ventral limbs did not (Figure 17.2). These results can be explained by assuming that the cells of the anterior–dorsal quadrant are resistant to changes in their positional memory by retinoic acid (Ludolph et al., 1990; Stocum, 1991).

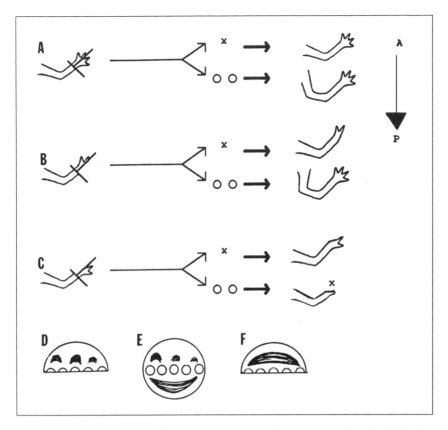

Figure 17.2 Effects of retinoic acid administration on normal and half limbs amputated through the distal radius and ulna. A: Normal regeneration without (x) or with (oo) retinoic acid treatment. The treatment results in proximalization. B: Regeneration of anterior half limb without and with retinoic acid treatment, respectively. C: Regeneration of half posterior limb without and with retinoic acid, respectively. D and E: Regeneration of control and retinoic-acid-treated half dorsal limb, respectively. Blocks indicate dorsal muscles and bands indicate ventral muscles. F: Regeneration of control half ventral limb. Half ventral limbs treated with retinoic acid fail to regenerate. (Adapted from Stocum, 1991.)

17.1 Presence of vitamin A metabolites in the blastema

The presence of endogenous retinoic acid has been verified in the regeneration blastema of *Amblystoma mexicanum* by means of high-performance liquid chromatography. The measurements were taken from a 14-day-old blastema. It was found that the concentration of retinoic acid is 5 times higher in the posterior than in the anterior edge. Interestingly, another gradient was observed along the proximal–distal axis. The concentration in the ulna-radius level was 2.5 times more than in the humerus level (Table 17.1) (Scadding and Maden, 1994). A comparative study in the frog *Xenopus laevis*, where regeneration does not result in regulated pattern, showed that no such gradients exist. The initial experiment where proximalization was observed suggested that retinoic acid, if present endogenously, should form a gradient with more present in the proximal regions than in the distal. The HPLC study, however, showed the opposite. It is possible that this graded distribution of retinoic acid is dependent on the stage of blastema formation; in earlier stages, it could be distributed differently. Usually, the proximalizing effects of retinoic acid have been determined when it was administered prior to day 14.

Brockes attacked this problem with a different methodology. He transfected newt limb cells with a reported gene whose expression is dependent on elements that respond to retinoic acid. He then implanted these cells in the amputated proximal or distal regions of the limb. He observed a 3.5 times increase in induction of the reporter gene in the proximal region of the limb, a result that indicates higher concentration in the proximal region; this result is in good agreement with the dose-dependent ability of retinoic acid to proximalize (Brockes, 1992). Such results add to the complexity of the problem. For example, the gradients could be "fluid." The distribution of vitamin A gradients could change constantly with time. This fluidity could, in fact, be affected by the presence of more than one metabolite capable of affecting the pattern (i.e., 9-*cis*-retinoic acid) and by the differential use of the receptors (see Section 17.2.2).

The ability of retinoic acid to duplicate structures during chick limb development and its endogenous presence as a gradient in the developing chick limb provided further strength to the idea of gradients as messengers of positional information (see the Addendum, Section 17.4).

17.2 Mode of action of retinoic acid

The different metabolites of vitamin A, such as retinol or retinoic acid, exert their effects by binding to two different proteins: (1) a carrier, the cellular retinoic acid binding protein (CRABP), and (2) a nuclear pro-

Table 17.1 Concentrations of retinoids in axolotl, radius-ulna-level limb regeneration blastemas divided in anterior to posterior quarters.

	Anterior quarter	Second quarter	Third quarter	Posterior quarter
Retinoic Acid	7.1 [53.6]	11.1 [88.6]	81.4 [661]	146.0 [1267]
	15.7 [178]	N/A [N/A]	55.1 [451]	75.6 [744]
	26.6 [244]	60.9 [460]	63.3 [479]	46.6 [386]
Mean	16.5 ± 10	36.0 ± 35	66.6 ± 13	89.4 ± 51
	[158 ± 97]	[274 ± 263]	[530 ± 114]	[799 ± 443]
Retinol-means of three measurements	337 ± 166 [2857 ± 913]	498 ± 57 [4080 ± 902]	457 ± 38 [3463 ± 155]	718 ± 448 [6282 ± 4102]
3,4-Didehydroretinol-means of three measurements	243 ± 214 [2139 ± 1863]	554 ± 161 [4541 ± 1622]	666 ± 213 [5073 ± 1790]	202 ± 15 [1567 ± 347]

Note: Unbracketed figures express the results in nanomolar concentration. Bracketed figures express results calculated as picograms of retinoid per milligram of protein. The means in each case are followed by the standard deviation (mean ± standard deviation). N/A (Not available) means the data were missing due to technical problems. Analysis of linear regression indicated that the results are statistically significant (r = 0.75; $P < 0.01$).

tein, termed retinoic acid receptor (RAR). A different receptor, RXR (retinoic X receptor), was termed because at the time of its discovery, its ligand (9-*cis*-retinoic acid) was not known. CRABP belongs to the family of transport proteins for small hydrophobic molecules, which includes such carriers as beta-lactoglobulin, apolipoprotein D, alpha-1-microglobulin, and BG protein from olfactory epithelium (Sherman, Lloyd, and Chytil, 1987; Godovac-Zimmerman, 1988; Giguere et al., 1990). RAR belongs to the superfamily of steroid receptors, proteins that are transcription factors, mediating gene activation by binding to DNA through specific domains, namely, the zinc fingers (Giguere et al., 1987; Giguere, 1994).

17.2.1 Structure of CRABP; binding of the ligand and expression

Initial studies, especially those based on sequence alignments of the transport proteins (already discussed), indicate that there are two stretches of sequences, approximately 100 amino acids apart, that are strongly conserved in all these proteins. The two sequences are the Gly-X-Trp-(His, Tyr) and Thr-Asp-Tyr. It was hypothesized that these two sequences form the ligand-binding domain. In the three-dimensional structure of β-lactoglobulin, these residues are located in the same region (Figure 17.3) (Godovac-Zimmerman, 1988). X-ray structure determination of the epididymal CRABP, which binds both all-*trans*-RA and 9-*cis*-RA, has provided supporting proof for such interaction. The structure shows a beta-barrel with an amphipathic cavity forming the binding site and the binding-site entrance. The deepest part of the cavity is composed of hydrophobic amino acids; charged amino acids are found more proximal to the entrance. In fact, the stretches of conserved amino acids point into the barrel and form the ligand-binding sites (Newcomer, 1993) (Figure 17.4, *color insert*).

To date, two isoforms of the CRABP are known. In mice, those two isoforms share 73 percent identity. These proteins are expressed in embryonic tissues including limbs, and they are induced at least fiftyfold upon treatment with retinoic acid (Sherman et al., 1987; Giguere et al., 1990). During chick limb development, the expression of CRABP forms a gradient that is opposite to the endogenous retinoic acid distribution (Maden et al., 1988). There is an excess in the anterior portion when compared with the posterior. The CRABP I clone has been iso-

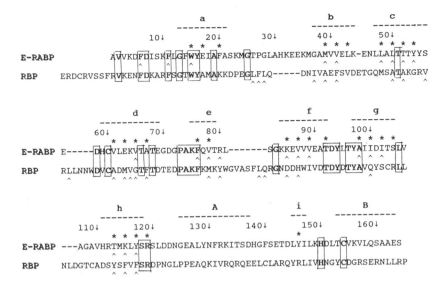

Figure 17.3 A structure based on sequence alignment of the amino acid sequences of E-RABP (top line) and retinol-binding protein (RPB, bottom line). The numbering is for E-RABP. Conserved amino acids are in bold and enclosed in boxes. The secondary structural elements of E-RABP are indicated: a to i are beta sheets, A and B are alpha helices. Beta-strand amino acids that point into the E-RABP barrel are marked with an asterisk. Amino acids that form the ligand sites are indicated with ^. (Courtesy M. E. Newcomer.)

lated from the axolotl embryo, but it was not expressed during limb blastema formation (Ludolph et al., 1993). This result could imply that the CRABP II could be involved in the process or that the CRABP I molecule might not be obligatory for the effects of retinoic acid.

Distribution of CRABP by means of protein isolation and binding activity of all-*trans*-RA or 13-*cis*-RA has been studied during limb regeneration in the axolotl. The level of CRABP in the cone blastema stage was found to be significantly higher than in the nonregenerating limbs. The level gradually falls to normal as regeneration is completed (Keeble and Maden, 1986). However, the level at the start of regeneration was not increased by pretreatment with a proximalizing dose of retinoic acid. The general idea about the role of CRABP is that it carries vitamin A metabolites to the nucleus, where they are able to bind to the receptors. However, interactions of CRABP with retinoids does not seem to be an obligatory step for the induction of proximalization during limb regeneration. Maden, Darmon, and Erikson (1990) and Maden et al. (1991) have synthesized analogues that do not bind to CRABP, and have shown that these analogues were still able to induce proximalization, even though they are less potent in this action than the ones that were able to bind to CRABP. The mechanism of transfer of retinoids to the nuclear RAR by CRABP is also unclear. Some scientists have shown nuclear localization of CRABP. Interestingly, CRABP shows homologies to nuclear proteins as well, especially to regions of leucine zippers (Tsonis and Goetinck, 1988).

17.2.2 Expression and structure of RARs and RXRs

The nuclear receptor for retinoic acid belongs to the superfamily of thyroid/steroid hormone receptors, which are relatives to the oncogene erb-A (Evans, 1988). These proteins contain primarily three domains. The N-terminal domain is associated with activity of the receptor. Next to that is the DNA-binding region of the receptor. This region is characterized by the presence of zinc fingers. A zinc finger is a sequence of about 30 amino acids with four cysteines spaced in a particular way. This arrangement is homologous to transcriptional factor IIIA and to other structures known to bind zinc. The four cysteines (in TFIIIA, two of them are substituted by histidines) are the ligands for the zinc binding. The characteristic three-dimensional structure of zinc fingers has been elucidated by NMR or X-ray diffraction studies. The 30 amino acids create an antiparallel beta-strand and a helix, which is the DNA-binding domain (Lee et al., 1993). Usually, each receptor contains two zinc fingers. The receptors dimerize in order to bind to DNA, forming specific sites at promoter regions that constitute the hormone-responsive elements (HRE). NMR studies of the three-dimensional structure of the DNA-binding domain of one receptor, RXR-alpha, indicate that an additional helix is present after the helix of the second finger (Figure 17.5). This is a unique feature and is in contrast to the DNA-binding domains of the glucocorticoid receptor and estrogen receptor (Lee et al., 1993). This additional helix defines a motif needed for selective dimerization of RXR with itself or with other members of the nuclear hormone receptor superfamily.

The ligand-binding domain lies in the C-terminus of the protein. Even though we do not have NMR or X-ray data on the ligand site for

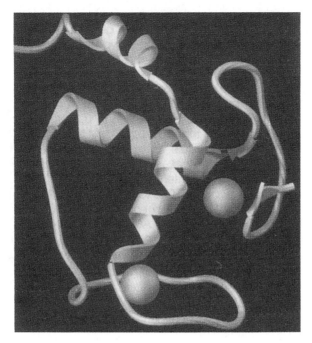

Figure 17.5 Structure of the RXR DNA-binding domain showing the packing of the three helices. The N-terminus is at right center, the C-terminus is at the upper left. Helix 1 extends from right to left across the center, helix 2 extends from bottom to top, and helix 3 is located at the top of the figure. The zinc atoms are represented as spheres. (Courtesy R. M. Evans.)

the receptors, some analogies can be made with the ligand site of CRABP. The sequences Gly-Leu-Trp and Thr-Asp-Leu are very similar to the two stretches of sequences in CRABP that are involved in the binding of retinoic acid (Figure 17.3). Interestingly, these two stretches are separated by almost the same distance in CRABP and RARs.

Studies in the mouse and in humans have identified three genes coding for the retinoic acid receptors: RAR-alpha, RAR-beta, and RAR-gamma. Comparison of their sequence shows divergence mostly in the N-terminus. Mouse studies have shown that RAR-alpha is ubiquitous during development, while the other two receptors show regulation in space and time. Another receptor, termed RXR, was found to be different from the RARs and is the receptor for 9-*cis*-RA (Kliewer et al., 1992; Levin et al., 1992). In newts, several of these isoforms have been cloned and their expression studied. The major isoform expressed by the blastema cells is RAR-delta1 (two receptors, delta1a and delta1b). The newt delta1 is the homologue of mammalian gamma1. This receptor is distinguished by a very unique N-terminus with a transcriptional activation domain unique among RARs (Ragsdale, Gates, and Brockes, 1992; Hill, Ragsdale, and Brockes, 1993). RAR-delta2 has also been cloned from the newt blastema. RAR-delta1 is expressed in the newt limb and the blastema. It is interesting, however, that not all cells in the limb tissues or the blastema express this receptor (Figures 17.6 and 17.7, *color insert*).

Two isoforms of RAR-alpha have been isolated (Ragsdale et al., 1989, 1992). The alpha1 is expressed in the limb, whereas the alpha2 is not. Partial sequences of a newt RAR-beta have also been reported (Giguere et al., 1989). The multiple forms of RARs could implicate each one of them in the different effects of retinoids on limb morphogenesis (Ragsdale et al., 1989). In fact, by the use of chimeric retinoic acid/thyroid hormone receptors, it was found that RAR-alpha1 mediates the growth inhibition observed by treatment with retinoic acid (Schilthuis, Gann, and Brockes, 1993). These chimeric receptors were transfected into blastema cells along with a marker plasmid expressing lacZ. The cells were also labeled with ^{3}H-thymidine. Growth of cells transfected by the alpha1 chimera was inhibited by treatment with T3 as judged by the incorporation of ^{3}H-thymidine. Cell growth was not affected when delta1 chimera was used. The action of retinoic acid was studied when these chimeric receptors were transfected into the wound epithelium *in vivo* by biolistic particle delivery system. These chimeric receptors were transfected with an alkaline phosphatase marker gene, activated with T3, and the expression of the marker and the retinoic acid–inducible WE3 antigen was analyzed. It was found that in cells transfected with the chimera delta1, many double-labeled cells were present, whereas limbs transfected with the chimaeric alpha1 had no double-labeled cells (Figure 17.8). These results indicate that different actions of retinoic acid can be mediated by different receptors (Pecorino, Lo, and Brockes, 1994).

Other studies have also implicated RXRs in limb regeneration. It was found that 9-*cis*-RA proximalizes limb regeneration in axolotls and is, in fact, more potent than all-*trans*-RA (Tsonis et al., 1994) (Table 17.2 and Figure 17.1). The same is true for chick limb development, where 9-*cis*-RA has been found to be as much as 25 times more potent than all-*trans*-RA (Thaller, Hofmann, and Eichele, 1993). RXRs have

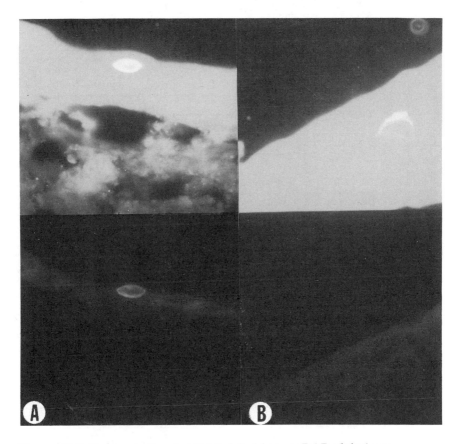

Figure 17.8 Demonstration that RAR delta1 but not RAR-alpha1 can increase WE3 reactivity in response to thyroid hormone. Limbs were transfected with an alkaline phosphatase marker plasmid that contained either chimeric alpha1 or chimeric delta1. The chimeric receptors were made in such a way that the ligand part of the RARs was substituted with the ligand part of the thyroid hormone receptor. These plasmids were transfected into intact limbs with biolistic transfection. A: Section through a limb transfected with chimeric delta1 receptor and treated with T3 4 days after transfection, showing a stained cell that is expressing alkaline phosphatase (upper panel) and WE3 (lower panel). B: Same experiment but with chimeric alpha1 transfection. Note that the WE3 antigen is not present, indicating that such expression is driven by a distinct RAR receptor. Bar 0.05 mm. (Courtesy J. P. Brockes.)

not yet been isolated from urodeles, but expression has been shown during chick limb bud development (Thaller et al., 1993). The ability of RXR to heterodimerize also with RAR, thyroid hormone receptor (TR), or vitamin D receptor (VDR) (Laudet and Stehelin, 1992) adds to the

Table 17.2 Effects of vitamin D on axolotl limb regeneration via 1P injection (up) and local bead implantation (below).

Treatment	Total # limbs	Proximalized (%)	Abnormal (%)	Inhibited (%)	Normal (%)	Total # affected (%)
Control	39	0(0.0)	1(2.6)	0(0.0)	38(97.4)	1/39(2.6)
9-*cis* (100µg)	146	55(37.7)	31(21.2)	55(37.7)	5(3.4)	141/146(96.6)
9-*cis* (50µg)	100	29(29.0)	47(47.0)	0(0.0)	24(24.0)	76/100(76.0)
1/4 9-*cis* (25µg)	61	0(0.0)	11(19.0)	0(0.0)	50(81.0)	11/61(19.0)
RA (100µg)	52	28(53.8)	13(25.0)	8(15.4)	3(5.8)	49/52(94.2)
1/2 RA (50µg)	64	0(0.0)	6(9.4)	0(0.0)	58(90.6)	6/64(9.4)
1/4 9-*cis*+1/2 RA	68	59(86.8)	7(10.3)	0(0.0)	2(2.9)	66/68(97.1)
1,25	12	0(0.0)	8(75.0)	0(0.0)	4(25.0)	8/12(75.0)
1,25+1/4 9-*cis*	41	0(0.0)	25(61.0)	0(0.0)	16(39.0)	25/41(61.0)
1,25+1/2 RA	7	1(14.3)	1(14.3)	0(0.0)	5(71.4)	2/7(28.6)
24,25	13	0(0.0)	8(61.5)	0(0.0)	5(38.5)	8/13(61.5)
24,25+1/4 9-*cis*	49	0(0.0)	28(57.1)	3(6.1)	18(36.8)	31/49(63.2)
24,25+1/2 RA	6	1(16.7)	0(0.0)	2(33.3)	3(50.0)	3/6(50.0)

Treatment	Total # limbs	Proximalized (%)	Abnormal (%)	Inhibited (%)	Normal (%)	Total # affected (%)
1,25	84	0(0.0)	30(35.7)	0(0.0)	54(64.3)	30/84(35.7)
1,25+1/4 9-*cis* RA	20	0(0.0)	10(50.0)	0(0.0)	10(50.0)	10/20(50.0)
1,25+1/2 RA	100	4(4.0)	37(37.0)	4(4.0)	55(55.0)	45/100(45.0)
24,25	82	0(0.0)	24(29.3)	0(0.0)	58(70.7)	24/84(29.3)
24,25+1/2 RA	23	0(0.0)	10(43.5)	0(0.0)	13(56.5)	10/23(43.5)
KH1060	34	0(0.0)	8(23.5)	0(0.0)	26(76.5)	8/34(23.5)
KH1060+1/4 9-*cis*	40	0(0.0)	13(32.5)	6(15.0)	21(52.5)	19/40(47.5)
Control Bead	30	0(0)	0(0.0)	0(0.0)	30(100)	0/30(0.0)

RA: all-trans-retinoic acid; 9-cis: 9-cis-retinoic acid; 1,25: 1,25 dihydroxyvitamin D3; 24,24: a metabolite of 1,25; KH1060: a potent vitamin D metabolite. Injections are 100µg/gr body(full dose). Beads were soaked in 25mg/ml solution.

complication of gene regulation and function of these receptors. These different homodimers and heterodimers recognize distinct elements in gene promoters and, therefore, one formation or the other would lead to specific gene expression. For example, it has been shown in other systems that vitamin D would promote VDR–RXR heterodimerization and up-regulation of vitamin D responsive genes. Addition of 9-*cis*-RA would inhibit this action of vitamin D by promoting RXR–RXR homodimer formation and down-regulation of vitamin D responsive genes

(MacDonald et al., 1993). During chick limb bud chondrogenesis, 9-*cis*-RA seems to antagonize the action of vitamin D as well (Tsonis and Sargent, 1996). Thyroid hormone alone does not affect limb patterns (Hay, 1956), but when it is co-injected with retinoic acid, the resulting regenerates have more proximal structures than the ones receiving retinoic acid alone. Anteroposterior or dorsoventral duplication of limb structures also resulted from the mixed dose of TH and RA (Vincenti and Crawford, 1993). Limited interaction during limb morphogenesis can also be seen when vitamin D and retinoic acid are combined. Interestingly, the highest percentage of proximalizations was seen when nonproximalizing doses of 9-*cis*-retinoic acid and all-*trans*-retinoic acid were combined (Washabaugh and Tsonis, 1995). Such results suggest that different steroid hormones or retinoids may function through similar or competitive pathways during limb patterning.

These properties of certain hormones and their receptors might indicate a more elaborate role of the different retinoids in limb morphogenesis. For example, it could be that the presence of all these metabolites and the use of different receptors are necessary for the retinoids to have an effect. Moreover, it could be the use of the receptors that determines the "gradient" distribution of retinoids in the developing and regenerating limb. The gradients observed by Scadding and Maden (1994) in the regenerating blastema were organized at the posterior–anterior axis and the proximal–distal axis at day 14 post-amputation. Intact limbs do not have such gradients. Therefore, the gradient is organized in response to blastema formation. Blastema formation, however, implicates the expression of genes, including those coding for the retinoid receptors. And, most likely, the proximalizing effects of retinoids can be elicited by the recruitment of more than one retinoid receptor. The use of the different receptors or reporter genes coupled to specific HRE can help answer these questions. With the use of cultures or *in vivo* transfections, the expression and the action of the different receptors can be studied. The chimeric receptors mentioned previously can help elucidate the roles of the different receptors. Other experiments could involve the use of reporter genes coupled with HRE specific for each homodimer or heterodimer. The expression of the reporter gene could then indicate the expression of the receptors and the formation of the different dimers. It could be that the mechanisms of pattern formation are much more complicated than can be explained by the simple existence of a gradient or coordinates.

17.2.3 Dedifferentiation and retinoic acid

As mentioned previously, the degree of dedifferentiation changes along the proximal–distal axis. More dedifferentiation takes place at a proximal amputation plane than at a distal plane (Ju and Kim, 1994). This correlation prompted Ju and Kim to examine the effects of retinoic acid on the extent of dedifferentiation at a distal amputation plane. It was found that a distal stump treated with retinoic acid exhibits a higher degree of dedifferentiation, virtually comparable to that of an untreated proximal stump. This finding links the effects of retinoic acid on patterning with the degree of dedifferentiation. In turn, this may indicate that the necessary patterning mechanisms are specified during the dedifferentiation process. Maden and Keeble (1987) have concluded that such extensive dedifferentiation results in lumps of fibronectin-positive extracellular matrix that are released into the blastema. Dedifferentiation of cartilage does not seem to affect pattern respecification induced by retinoic acid because removal of all cartilage does not interfere with the proximalizing effects of retinoic acid.

17.2.4 Gradients revisited

If gradients specify positional information, they should be present or produced endogenously after amputation. The source of retinoids is not well understood. But when a limb is proximalized it contains two regions that specify the same positional information, two regions that will regenerate the same distal structures if amputated. If the morphogen is organized endogenously as a gradient, how can it be at the same concentration in two places? What is common in those two places is the same type of cells as far as their position is concerned. For example, on a proximalized limb we can have two cells with identical positions, say at the mid-ulna radius level. These cells respond to the environment in the same way because they are identical in gene activity, including receptors for retinoids. These cells will use retinoids in an identical fashion. Therefore, the *use* of retinoids might create the gradient and not vice versa. The distribution of the carrier protein CRABP is in fact opposite to the endogenous distribution of retinoic acid (Maden et al., 1988). This could level the use of retinoic acid across the anterior–posterior region.

Can such an idea reconcile the PCM theory with the theory of gradients? Work with the regeneration of double anterior limbs has suggested that the spacing of the positional values is not equal in the posterior and the anterior halves; more of them should be carried in the posterior half. We also know now that retinoic acid is more concentrated in the posterior region. This is the first coincidence of PCM with gradients. Similarly, those experiments suggest that this distribution of the circumferential positional values will be more equal in the shank (distal) than in the thigh (proximal). It is assumed that the levels of retinoic acid will be higher in the proximal than in the distal (based on the dose-dependent effects of retinoic acid on proximal–distal duplications). The same is suggested by Brockes (1982), who found that cells carrying a reporter gene under the control of a retinoid response element showed more expression when they were implanted in the proximal versus the distal region of the limbs. The distribution, however, of retinoic acid, as measured by HPLC (Scadding and Maden, 1994) is higher in the distal than in the proximal region. If retinoids could specify positional values as portrayed by the PCM, it would mean that the spacing is affected by the concentration of retinoic acid (less spacing or more values equals more retinoic acid). Specific gene expression can then occur according to the concentration of retinoic acid and the differential use of receptors. Such specific expression of the *wingless* gene has been shown in the imaginal disks of *Drosophila*, where it was found that *wingless* was expressed only within the boundaries of one angular positional value (Marsh and Theisen, 1993). The ability of Sonic hedgehog (*Shh*) (see Chapter 18.2.7) to act as determinant of the skeletal elements during limb development (Laufer et al., 1994) has been received as an important component of the morphogen idea. In fact, it has now been found that *hh* can signal at short range (locally) and at long range (graded) (Johnson and Tabin, 1995). The different signaling abilities of *hh* can be specified by the self-cleaving mechanism of the protein. This action of *hh* provides the ground for reconciling the different theories. It is my belief that both local cell-to-cell interactions and gradients in combination are necessary for morphogenesis.

17.3 Genes affected by vitamin A

Vitamin A compounds such as retinoic acid have been found to affect the expression of many other genes. Therefore, identification of such genes expressed in limbs could give valuable data on regulation by retinoic acid. If, for example, a gene is regulated by retinoic acid in a dose-dependent fashion, then regulation along the proximal–distal axis of the limbs could provide evidence of a graded distribution of retinoic acid.

Keratins were considered good candidates for such regulation. Keratins are intermediate filaments and their expression has been regarded as a marker for epithelia and epidermal differentiation. Furthermore, their expression is regulated by retinoic acid, which is a modulator of epithelial differentiation. The interest in keratins as markers for the action of retinoic acid in blastema formation arose from the study by Ferretti et al. (1989), where it was shown that certain keratins were expressed by the mesenchymal cells of the blastema (Figure 17.9). Subsequently, Ferretti, Brockes, and Brown (1991) cloned the cDNA of a type II keratin from the newt whose expression is restricted to normal and regenerating limbs and tails. More interestingly, the expression of this keratin was down-regulated by retinoic acid. Its level of

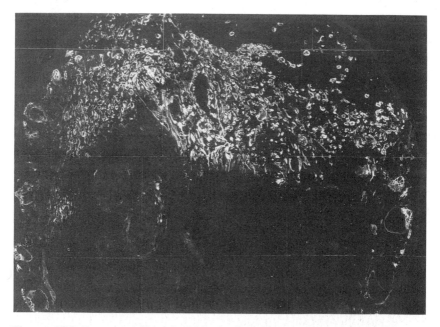

Figure 17.9 Presence of keratin 18 in the limb blastema, shown by staining of a blastema section with an antibody to keratin 18. Note the reaction only in the blastema and the glands. The wound epithelium and the intact skin are not positive. This keratin seems to be specific for the blastema. (Courtesy P. Ferretti.)

expression was higher in a distal normal limb and blastema and lower in a proximal normal limb and blastema (Figure 17.10). This observation indirectly supported the idea of the presence of more retinoic acid in the proximal region of the limb and implicates the presence of a proximal–distal gradient. However, more detailed analysis on the expression of this type II keratin (NvKII) showed that it is expressed only in the distal parts of the epithilium of the regenerating digits (Figure 17.11) (Ferretti, Corcoran, and Ghosh, 1993). Such a property might indicate that there is no proximal–distal gradient in the normal limb, but instead

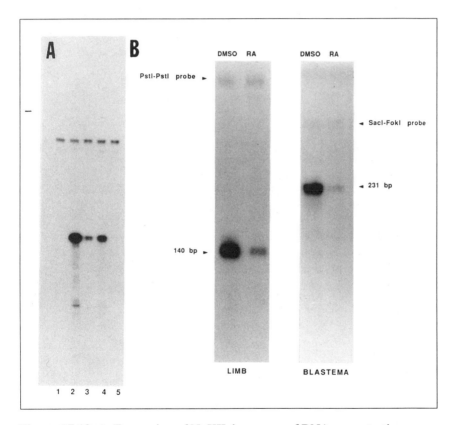

Figure 17.10 A: Expression of NvKII, by means of RNAase protection assay indicating elevated expression of this keratin transcript in the distal part of the limb relative to proximal ones. Lane 1: proximal normal limb; lane 2: distal normal limb; lane 3: proximal blastema; lane 4: distal blastema; lane 5: tRNA. B: Down-regulation of NvKII transcript after treatment of the limbs and blastema with retinoic acid (RA). (Courtesy P. Ferretti.)

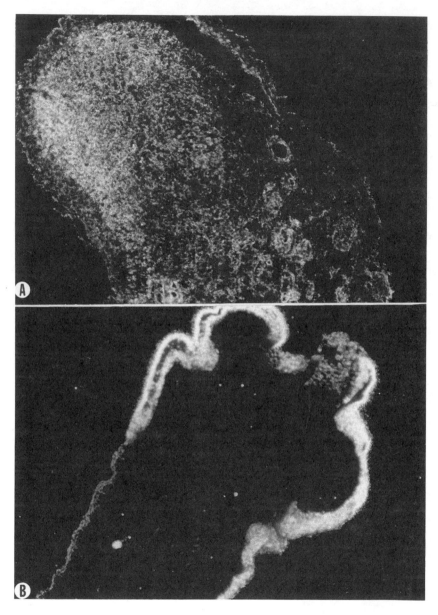

Figure 17.11 *In situ* hybridizations indicate expression of keratin transcripts during limb regeneration in *N. viridescens.* A: Antisense Nvk8 probe hybridization showing strong hybridization in the blastema mesenchyme, but not in the epidermis or the wound epithelium. B: Antisense NvKII probe hybridization showing hybridization in the epithelium of regenerating digits only. (Courtesy P. Ferretti.)

that the tip of the fingers is blastema-like and expresses blastema antigens (Ferretti, unpublished observations).

The role of certain glycoproteins has been suggested in the proximalizing effects of retinoic acid on the axolotl regenerating limbs. This was inferred from the effects of tunicamycin on the proximalization induced by retinoic acid. Tunicamycin is a drug that blocks the N-glycosylation of proteins. When used in conjunction with a proximalizing dose of retinoic acid, the effect of retinoic acid on limb patterns was abolished (Johnson and Scadding, 1992). This implicates asparagine-linked glycoproteins in the pathway through which retinoic acid induces proximalizing effects. Similarly, tunicamycin interferes with the ability of retinoic acid to induce duplications in the developing chick limb (Johnson and Langille, 1993)

Other genes have been sought as markers along the proximal–distal axis. Homeo-box-containing genes are under investigation in many different laboratories. These genes have an established role in the determination of cell lineages, differentiation, and pattern formation. These genes are also susceptible to regulation by retinoic acid. An initial report by Tsonis and Adamson (1986) and subsequent ones by Stornaiuolo et al. (1990) and Simeone et al. (1991) have conclusively shown that certain Hox genes are regulated by retinoic acid. In addition, expression of such genes has been shown during blastema formation and regeneration. Chapter 18 will examine in more detail the structure, function, and role of Hox genes in limb regeneration.

17.4 Addendum

17.4.1 Effects of retinoic acid on patterning in the developing limb

Transplantation experiments have clearly demonstrated that during early chick limb bud formation (stage 20–21), a region of the bud is able to specify skeletal development. This region, found in the posterior part of the limb bud, is called the zone of polarizing activity (ZPA). Transplantation of ZPA cells into the anterior margin of the limb bud results in duplication of the skeletal elements along the anterior–posterior axis (Saunders and Gasseling, 1968) (Figure 17.12). Vitamin A has been found to alter the patterns during chick limb development in a similar fashion to ZPA cells. In the initial experiment by Tickle et al. (1982), retinoic acid was inserted in the anterior margin, and duplication of the digits resulted (Figure 17.12). This effect, as in the case of the regenerating limb, is dose-dependent. Experiments with analysis of endogenous retinoids showed that the concentration of retinoic acid was about 3 times higher in the posterior than in the anterior margin of the chick limb bud (Thaller and Eichele, 1987). This indicated that retinoic acid might exist as a gradient in the limb bud, and, by that property, it might specify positional information. An antibody to CRABP, however, showed an opposite distribution of the carrier protein in the chick limb (Maden et al. 1988). Such a distribution, however, was not observed in mouse limb buds (Dolle et al., 1989a; Perez-Castro et al., 1989). Distribution of RARs has not revealed any particular graded expression. These observations just add to the mystery of retinoic acid's mode of action. The opposite dis-

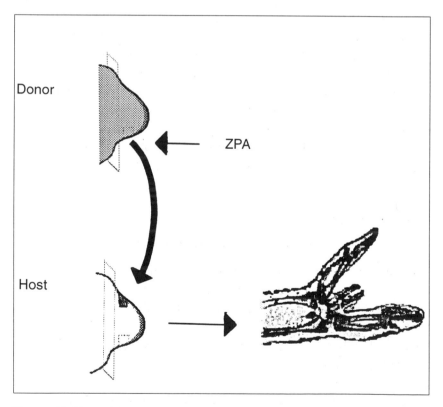

Figure 17.12 An illustration depicting the effects of ZPA transplantation or retinoic acid administration onto the anterior region of a developing chick limb bud. (Adapted from Hofmann and Eichele, 1994.)

tribution of retinoic acid and its carrier protein suggest equal use in both anterior and posterior margins. On the other hand, CRABP has been shown not to be necessary for the morphogenetic effects of retinoic acid during limb regeneration (Maden et al., 1990, 1991).

A different explanation of the mode of action of vitamin A has been proposed by Wanek et al. (1991). These investigators placed a bead with retinoic acid in the anterior margin of a developing limb bud. Several hours later (after the retinoic acid had been released, degraded, and had produced its effects), the investigators removed the tissue neighboring the bead implant and implanted it into the anterior margin of another host. Duplication resulted. The authors argue that retinoic acid induced the differentiation of the anterior cells to become posterior cells and that it is not the graded distribution of the agent that is responsible for the morphogenetic effects. Support for such a conclusion comes from molecular data received by Noji et al. (1991). These researchers showed that when the developing chick limb bud is treated with a dose of retinoic acid that induces limb duplications, the expression of RAR-beta is induced to a much higher level than that in the posterior polarizing region. This implies that the ZPA normally contains less retinoic acid than is necessary for limb duplication and that retinoic acid might not be the endogenous signal. In discussing this topic further, Hofmann and Eichele (1994) maintain that when retinoic acid is applied at appropriate doses, it does not directly induce anterior cells to become posterior cells. Instead, it induces expression of posterior-specific Hox genes that eventually transform the anterior cells into a posterior character. The suppression of 5' Hox genes by retinoic acid (which show expression at the distal regions) adds to the argument against its role in specifying pattern formation (Stornaiuolo et al., 1990; Simeone et al., 1991). In fact, it seems that there are mechanisms during embryo development that ensure no expression of retinoic-acid–responsive genes in the distal part of the limb bud.

In a transgenic mouse, made with lacZ driven by retinoic-acid–responsive elements (RARE), there is no expression of the lacZ in the distal limb buds. Expression could be observed in the proximal regions of the limb bud and in the distal after treatment with retinoic acid (Rossant et al., 1991). Such data suggest there exist mechanisms that ensure that distal retinoic acid is not functional. Such an idea is against the role of retinoic acid in patterning, because such patterns are set in the distal part of the limb bud. Instead, the data presented by Wanek et al. (1991) and Bryant and Gardiner (1992) suggest that retinoic acid converts cells to posterior–ventral–proximal and then these cells interact with adjacent anterior cells producing the skeletal duplications.

The isoform 9-*cis*-retinoic acid can also duplicate the skeletal elements of the digits (Figure 17.13) (Thaller et al., 1993; Tsonis et al., 1994). This compound was found to be about 25 times more potent in this action than all-*trans*-retinoic acid. In fact, Thaller et al. suggest that 9-*cis*-retinoic acid might be the active morphogenetic compound, since a 4 percent transition of all-*trans*-retinoic acid to 9-*cis*-retinoic acid was observed. 9-*cis*-retinoic acid binds to RXR as well as RAR. RXR was found to be expressed in the limb buds at the time of administration; therefore, this receptor might be involved in the morphogenesis of limbs.

17.4.2 The developing amphibian limb

The action of retinoic acid on the amphibian regenerating limb (proximalization) and on the chick developing limb (posteriorization) brings up the following question: What is the effect of retinoic acid on the developing amphibian limb? To answer this question, Scadding and Maden (1986) used axolotl larvae at stages where the limbs were in early (elongated bud) or late (two digit) developmental stages. By using

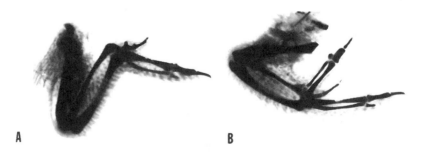

A   **B**

Figure 17.13 A: A normally developed chick wing. B: Duplication of wing structures after a bead soaked in 9-*cis*-retinoic acid was inserted in the anterior region of a developing wing at stage 21. Note the duplication (43234) along the anterior–posterior axis. The wings are stained with Victoria blue B.

such early stages, the authors were able to amputate the right limbs and leave the left to continue development. These methods allowed the investigators to compare the effects of vitamin A (retinol palmitate) on both the regenerating and the developing limbs in the same animal. The developing limbs responded to the treatment differently from the regenerating ones. Although the general effects of vitamin A compounds on limb regeneration (that is, proximalization) were confirmed with the regenerating limbs, the treatment caused developing limbs to become hypomorphic. The degree of skeletal element deletion varied depending on the dose and stage. Thus, when early stage limbs were treated, the deletions were greater. These results clearly indicate that vitamin A affects the developing and the regenerating limbs differently. The differential use of RARs has been thought to be of vital importance in the different effects of retinoic acid in these two systems. However, since the experiments were designed in the same animal, it is difficult to realize differential expression of RARs, unless amputation triggers different events even in the developing embryo. Muneoka and Bryant (1982), using basically the same experimental method, have shown that the patterning mechanisms in developing and regenerating limbs are virtually the same. If RARs are responsible for patterning, then these effects of retinoic acid add to the complexity of the problem. An explanation could be that the regenerative mechanisms (inductive) are different from the ones that induce limb bud formation in the embryo. This does not exclude the possibility that patterning mechanisms could be the same (see also Chapter 19).

Do these results mean that the amphibian limb does not have the equivalent of an avian ZPA? As already discussed, Harrison (1921) had observed that when a limb disk was grafted into the normal site in reversed anterior–posterior orientation, it often gave a reduplication; however if grafted in normal orientation, it gave a normal limb. When a limb disk was grafted into the flank with normal orientation, it produced reduplication! Such events were also reported by Slack (1976), who, in addition, found that the proximal flank contains polarizing activity. This could explain why reduplication occurs when a limb disk is transplanted into the flank with its normal axis retained. Slack believes that Harrison's experiments can be explained by the fact that polarizing activity exists in two sites after reversal of the axis and transplantation, making two posterior regions. Slack concluded that a posterior polarizing region exists in the amphibian limb and that, in this sense, amphibian limb development is like that of other vertebrates, such as birds. The presence of a ZPA in the developing amphibian limb would not necessarily contradict the differing effects of retinoic acid on the developing limbs of amphibians and chick. Instead, it indicates that variability might exist in the function of the receptors for retinoic acid or in the distribution of retinoids across the developing limb bud of amphibians.

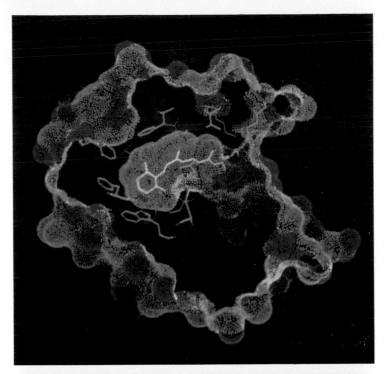
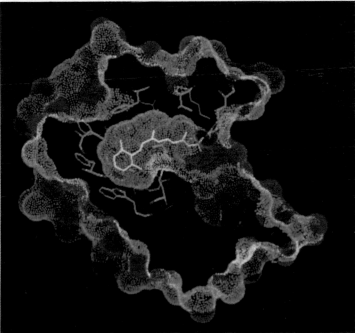

Figure 17.4 Stereo view of the binding cavity and binding site entrance of ependymal retinoic-acid-binding protein with bound ligand (in cyan). Atom coloring is as follows: C: green, N: blue, O: red, S: yellow. (Courtesy M. E. Newcomer.)

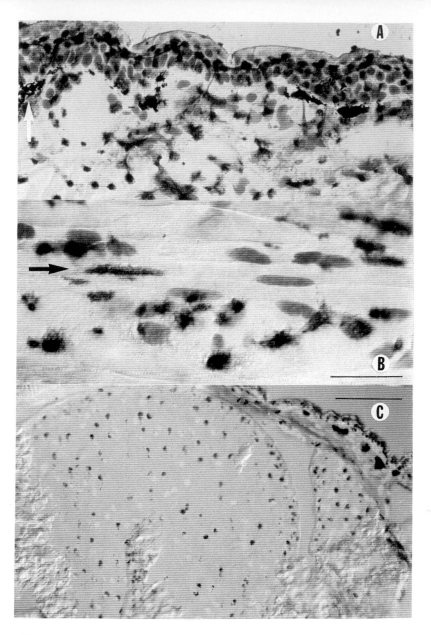

Figure 17.6 Expression of RAR-delta1 in the newt limb. A: Section through the skin and the underlying mesenchyme. The arrow indicates a chromatophore. B: Section through the skeletal muscle. C: Section through cartilage. Nuclei were counterstained with cresyl violet (A, B) and Hoechst dye (C). Note that positive and negative cells are intermixed in A and C. In B, positive and negative nuclei can be observed within the same fiber (arrow). This shows heterogeneity in the expression of this receptor in limb tissues. These sections were stained with an antibody made from the N-terminus of the protein. Bars 0.1, 0.05, and 0.2 mm in A, B, and C, respectively. (Courtesy J. P. Brockes.)

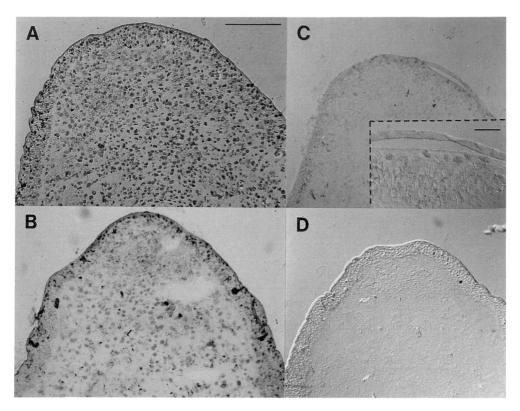

Figure 17.7 Expression of RAR-delta in sections from an 18-day-old blastema (late bud) from *N. viridescens*. A: Stained with an antibody made against the N-terminus (as in Figure 17.6). B: Stained with an antibody made against sequence of the C-terminus. In A and B, about 50 percent of the cells are positive. C: Stained with an antibody against a synthetic peptide made from sequences covering the delta1a and delta1b initiation sites. The intensity of staining is weaker than in A and B, but nuclei in the wound epidermis are strongly immunoreactive (insert). D: The negative control. Bar 0.045 mm. (Courtesy J. P. Brockes.)

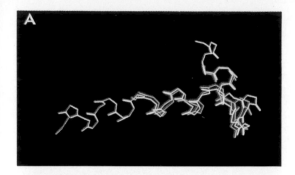

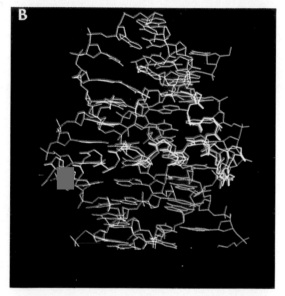

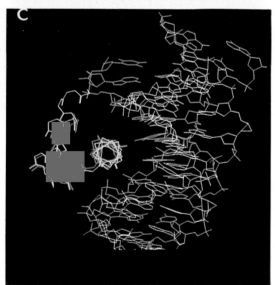

Figure 18.2 Three-dimensional structure of the helix-turn-helix (HTH) domain of the engrailed homeo domain and its comparison with the same domain from lambda repressor. A: View of the HTH domains roughly perpendicular to the recognition helices. Residues from the repressor are shown in orange and from the homeo domain in yellow. Note the similarities in structure and the elongated helix for the homeo domain. B and C: Repressor– operator and homoeodomain–DNA complexes seen from different perspectives. The operator is shown in blue and the engrailed binding site in purple. Note the difference in DNA bending. (Courtesy C. O. Pabo.)

18

Hox Genes and Limb Regeneration

The homeo box is a 180-nucleotide sequence that was initially found in three different homeotic genes in *Drosophila*. These genes, the *fushi tarazu, ultrabithorax,* and *antennapedia*, are responsible for specification of cell lineages and pattern formation during *Drosophila* development (Gehring, 1985). The amino acid sequence of the homeo box was found to be very similar to the helix-turn-helix motif of the bacterial repressors. This characteristic structure has the property of binding to DNA. Subsequently, homeo box sequences were found in all organisms tested. In numerous examples, these sequences are expressed in a manner that correlates with the laying down of patterns during development and the determination of cell lineages. The degree of conservation among homeo box sequences in different animals is astonishing. In many instances the homeo box of the same gene can be 100 percent identical among two distantly related animals.

The organization of Hox genes in the chromosomes is also interesting. There are several clusters of Hox genes, designated A, B, C, and D.

Each cluster contains several Hox genes designated A1, A2, ... These genes seem to be sequentially expressed with the gene at the 3' being first and the gene at the 5' being last (Acampora et al., 1990; Belleville et al., 1992). This pattern coincides with morphogenetic events along an axis. For example, genes of the D cluster that are located at the 3' end are expressed anteriorly during limb development, whereas genes at the 5' end are expressed more posteriorly (Figure 18.1). A similar situation exists in *Drosophila* development (Brockes, 1989; Dolle et al., 1989b; Duboule, 1991; Izpisúa-Belmonte et al., 1991). Further support that Hox genes specify positional information has come from experiments where misexpression of targeted Hox D 11 caused homeotic transformation in

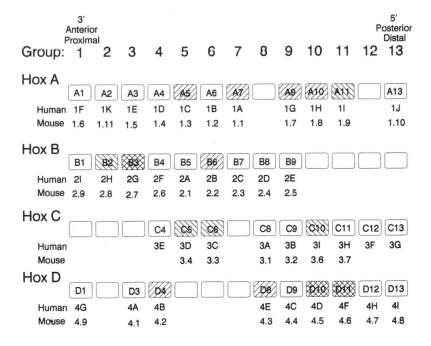

Figure 18.1 An illustration demonstrating the mapping of the different Hox genes in clusters. Hox genes bearing small numbers (at the 3' location on the chromosome) are usually expressed at the anterior and proximal regions of developing embryos and limbs. Hox genes located at the 5' region of the chromosome are usually expressed at the posterior and distal regions of the developing embryos and limbs. Some of them (shaded) have been cloned from newts. (Adapted from Hofmann and Eichele, 1994.)

the developing limb (Morgan et al., 1992). Targeting of Hox genes has clearly shown a proximal-to-distal effect. The most 5' ones (such as D13) affect the phalanges. A11 or D11 alone affects structures in the digits and ulna or radius, whereas D10 or C10 affects the humerus. Double mutants of A11 and D11 are missing ulna and radius. In other words, it seems that Hox13 genes (from all clusters) are all needed to create the hand, Hox 12 the carpals, Hox 11 ulna and radius, Hox 10 the humerus, and Hox 9 the scapula (Davis and Capecchi, 1994; Davis et al., 1995). Justifiably, Hox genes have been considered appropriate candidates for specifying positional information and cell differentiation during amphibian limb regeneration. This chapter will review the information available on this topic.

18.1 Structure of the homeo domain

As already mentioned, the homeo domain contains a sequence homologous to the helix-turn-helix motif of the bacterial repressors. NMR studies on the three-dimensional structure of the 60-amino-acid peptide that constitutes the homeo domain have indicated a structure of three helices. Helices 2 and 3 form the helix-turn-helix motif. Helix 1 is at the N-terminus of the peptide and, as a matter of fact, it contains amino acids quite conserved in homeo boxes and is homologous to helical domains found in other proteins (Nicosia et al., 1990). Helix 3 is longer than its bacterial repressor counterpart. The N-terminal part fits in the minor groove of DNA, and helix 3 makes contacts with bases of the major groove and is the recognition helix (Qian et al., 1989; Kissinger et al., 1990). The homeo domain differs from the bacterial repressors in the mode of dimerization (if any) and in the mode by which it binds to DNA. In the repressor–operator complex, helix 2 fits partway in the major groove. The corresponding helix of the homeo domain lies above the major groove. Superimposition of the two complexes indicates that the DNA for the homeo domain has been rotated away from the N-terminal of that helix (Kissinger et al., 1990) (Figure 18.2, *color insert*). The homeo domain of the hepatocyte-specific factor LFB1 is different from ordinary homeo domains in that it contains 21 extra amino acids between helix 2 and helix 3 and exists as a dimer (Nicosia et al., 1990). Dimerization seems to occur through the N-terminal sequences. In that region there are sequences that resemble sequences of myosin. These sequences fold to form alpha helices in the rod portion of myosin. The homeo domains might dimerize by adopting a similar conformation.

18.2 Expression of Hox genes in the regenerating limbs

Several homeo box–containing genes have been isolated from libraries made from limb or blastema RNA. These genes have in turn been used for the study of regulation of regeneration and patterning in the regenerating limb.

18.2.1 Limb-specific Hox genes

Initial studies by Tabin (1989) led to the isolation of two Hox genes that were expressed differentially in the newt forelimb and hindlimb. One of them (termed FH 1) was found to be expressed only in the forelimb. The expression was restricted to the epidermal cells only; therefore, this gene was not considered to play any role in patterning. A second one, termed FH 2, was isolated by Tabin and independently by Savard, Gates, and Brockes (1988) (termed NvHBox 1; homologous to mouse Hox-3.3). This gene is highly expressed in the unamputated forelimb and the proximal blastema. There is expression in the hindlimb, but it is much less than in the forelimb. Expression of NvHBox 1 was not observed in the adult *Xenopus* limbs. The newt Hox-3.6 was found to be hindlimb-specific, but it was also expressed in the blastema (Simon and Tabin, 1993). Two other Hox genes have been isolated and shown to be expressed in newt limbs. These genes contain a homeo box that is highly homologous to the homeo box of the *Drosophila* homeotic gene *Distal-less*. In the fruit fly this gene is responsible for the determination of the distal structures of the limb, and mutations affect the distal elements of the limb (Cohen et al., 1989). The newt counterparts of the *Distal-less* were both found to be expressed in unamputated limb skin (Beauchemin and Savard, 1991, 1992). Four Hox genes have been isolated from a newt genomic library. Those are the Hox-3.3, Hox-3.4, Hox-2.7, and Hox-2.8 genes. All except Hox-3.4 are expressed in unamputated adult limbs, even though the expression of their mammalian counterparts in limbs is restricted to their early development (Belleville et al., 1992). Transcription of Hox-3.3 (C6) produces two transcripts that differ in their 5' region. This gene is also expressed in the limbs of adult *Xenopus* (Savard and Tremblay, 1995). The difference between this result and the one that showed no expression in the *Xenopus* limbs (Savard et al., 1988) could be explained by the sensitivity of the method employed, with RNase protection producing positive expression and Northern analysis negative. Interestingly, the chromosomal organization of these newt Hox genes seems to be similar to that of other vertebrates. Despite the large C value in the newt genome, the intronic spacers separating Hox-3.3 from Hox-3.4, and Hox-2.7 from Hox-2.8, are similar to those of the homologous regions of the mouse and human (Belleville et al., 1992). The Hox A11 (1.9) cDNA has also been cloned and shown to be expressed in adult muscle, bone, and skin of the limb – tissues that contribute to regeneration. Similar expression in the adult limb has been shown for the HoxA 3 (2.7) cDNA clone (Beauchemin et al., 1994; Beauchemin, Tremblay, and Savard, 1995). The NvHBox 6 (the newt version of mouse and human emx-2) has also been found to be present in the skin of the intact limbs.

18.2.2 Blastema-specific Hox genes

Several Hox genes have been found to be specific to or up-regulated in the regeneration blastema. The NvHBox 2 (Hox-4.6) isolated by Brown and Brockes (1991) is specific for the regeneration blastema and is restricted to the mesenchymal cells only. Three others isolated by Beauchemin and Savard (1993) were found to be expressed in the intact limb and up-regulated during blastema formation. The NvHBox 2.7 and NvHBox 3 (homologous to Hox-2.7 and Hox-1.9, respectively) were found to be 10 times higher in the blastema than in the intact limb. The NvHBox 6 (homeo domain homologous to the one from the *Drosophila Empty spiracle*) was found to be skin specific and up regulated 5 times in the blastema (Beauchemin, Tremblay, and Savard, 1995). The newt Hox-4.5 (Simon and Tabin, 1993) was found to be specifically expressed in the blastema. Up-regulation during blastema formation has been shown for Hox A3, HoxA 11, and HoxC 6 (Beauchemin et al., 1994, 1995; Beauchemin, Tremblay, and Savard, 1995).

18.2.3 Position-specific expression of Hox genes

Some of the Hox genes that are up-regulated during blastema formation have been found to have position-specific expression along the different limb axes. NvHBox 1 and NvHBox 2 transcripts were more abundant (3–5 times) in blastema from the proximal compared to distal level. NvHBox 6 was found to be higher in the distal than the proximal level. Finally, an axolotl Hox gene, the AHox-4.5, was expressed with higher levels in the posterior than in the anterior blastema (Gardiner, Blumberg, and, Bryant, 1993; Gardiner et al., 1993). The newt homologue of Hox-4.5 was found to be expressed more in the proximal than in the distal blastema. One of the Hox C6 transcripts shows graded proximal–distal expression (Savard and Tremblay, 1995). Higher expression in the proximal versus the distal blastema has been demonstrated for HoxA 11 and HoxA 3. Interestingly, the NvHBox 6 transcript shows higher levels of expression in the distal than in the proximal (Beauchemin et al., 1994, 1995; Beauchemin, Tremblay, and Savard, 1995).

18.2.4 Other Hox genes

Nineteen more Hox genes have been cloned from axolotl limb libraries (Gardiner et al., 1993). Most of these genes were homologous to other Hox genes found to be expressed in the limbs of several species (the exception being the Hox-2.7 [B3] gene). These included A4, A5, A7, A9, A10, A11, B3, B5, B6, C13, D4, D8, D10, D11, and six unidentified ones. Detailed expression studies of these axolotl Hox genes have not been reported yet. The axolotl A13, which is at the most 5' location (Figure 18.1) has been found to be turned on as soon as 1 day post-amputation (distal or proximal amputation). The expression is restricted to the distal part of the stump before blastema formation. The same is true for A9, but it differs from the expression in the developing limb, where its expression is more proximal (Bryant, 1994; Gardiner et al., 1995). The homologues for msh (muscle-specific Hox) and engrailed were among the isolated ones as well. The newt *msx-1* has been isolated and shown to be expressed during the stages when undifferentiated blastema cells are present (Simon et al., 1994). The HoxA 2 and HoxA 3 have also been cloned from the newt; they have been isolated and shown to be expressed in the limb and blastema (Savard, personal communication).

18.2.5 Regulation of Hox expression by retinoic acid

Despite the observation that several newt Hox genes were expressed differentially according to position (along the proximal–distal axis), regulation by retinoic acid was observed only in the case of Hox-4.5 and NvHBox 6 (Simon and Tabin, 1993; Beauchemin, Tremblay, and Savard, 1995). One would expect that a proximalizing dose of retinoic acid would create distal expression levels similar to proximal ones. But this was the case only with the aforementioned Hox genes so far. This means that the Hox-4.5 and NvHBox 6 might be part of the pathways whereby retinoic acid exerts its morphogenetic effects during limb regeneration. The axolotl A13 has been found to be down-regulated by treatment with retinoic acid (Bryant, 1994). In fact, work from other systems has shown that retinoic acid up-regulates 3' Hox genes that are expressed in proximal and anterior regions, but it down-regulates the Hox genes that are expressed in distal and posterior regions (Stornaiuolo et al., 1990; Simeone et al., 1991). In this respect, Hox A1, B1, C1, and D1 are the most responsive to retinoic acid, whereas Hox A13, B13, C13, and D13 are inhibited by the treatment.

Table 18.1 Hox genes from *N. viridescens* in limb regeneration.

Name			Expression			
Old	New	Other	Intact Limb	blastema upregulation	proximal-distal gradient	regulation by retinoic acid
		FH1	+ (forelimb)			
3.3	C6	FH2, NvHBox-1	+	+	+	−
3.6	C10		+ (hindlimb)	+		
3.4	C5					
2.7	A3		+	+	+	−
2.8	B2		+			
1.9	A11		+	+	+	−
4.6	D11	NvHBox-2		+	+	−
4.5	D10			+	+	+
		msx-1		+		
		NvHBox-6	+	+	+ (distal)	+
		Distal-less 1	+			
		Distal-less 2	+			

A summary of all Hox genes isolated in *Notophthalmus viridescens* and related to limb regeneration is presented in Table 18.1.

18.2.6 Hox genes and homeotic transformation

As mentioned previously (Chapter 15), vitamin A can induce generation of limbs at the amputation site of tails in tadpoles (Mohanty-Hejmadi et al., 1992). Whether or not a homeo-box–containing gene is involved in such an effect of vitamin A is not yet known. However, some similarities exist with this type of homeotic transformation and others seen in *Drosophila* where Hox genes are involved. The transformation of an amputated antenna into a limb has been observed by Herbst in 1901. This is reminiscent of the *Antennapedia* mutation in *Drosophila* (Gehring, 1985). The involved gene contains a homeo box and historically was one of the first to be identified. These three facts – the spontaneous transformation, the induced transformation by vitamin A, and the mutation involving a homeo box – suggest that Hox genes might be involved in homeotic transformation of tails to limbs during regeneration.

18.2.7 Hox genes and limb regeneration in higher vertebrates

As mentioned in Chapter 1, the capacity for limb regeneration is mainly limited to some urodeles. Regeneration in higher vertebrates has been studied in mice and in clinical cases involving humans. It has been shown that mice can regrow the tip of their toes if amputation is performed distally to the first interphalangeal connection. Recent studies have involved the expression of the homeo-box–containing gene *msx*-1. This gene (*msx:* muscle-specific hox) has been found to be inducible by the apical ectodermal ridge (AER). Regeneration of the toe tip is correlated with expression of *msx*-1. When the amputation is permissive for regeneration, extensive expression can be observed in the toes of the mouse. This is not the case, however, when the amputation is not permissive for regeneration. The AER, therefore can be implicated in regeneration processes as well (Muneoka, 1993). The AER is known to be important for limb development. Studies during chick limb development have clearly shown that development is stopped if the AER is removed. It is believed that AER secretes growth factors that stimulate the growth and differentiation of the underlying mesenchyme. If the

AER is removed, but FGF4 or FGF2 is administered to the mesenchyme, distal patterns are generated (Niswander et al., 1993; Fallon et al., 1994). Because FGF4 is localized in the posterior ridge, it has been proposed that FGF4 is the ridge signal to mesenchyme.

Another molecule, the Sonic hedgehog (*Shh*), is localized in the area of ZPA and is believed to be the polarizing signal. It has recently been found that a positive feedback loop exists, with FGF4 and *Shh* the key players. Expression of FGF4 in the ridge can be regulated by *Shh*, and that *Shh* expression in the mesenchyme can be activated by FGF4 and retinoic acid (Niswander et al., 1994; Laufer et al., 1994). Similarly, FGF2 can induce regeneration in the amputated stage 25 chick limb bud and in additional limbs in the chick embryo when H is applied to the flank regions of the body (Cohn et al., 1995; Taylor et al., 1994). The working scenario here is that the stimulus is provided by retinoic acid and FGFs and that *Shh* and Hox genes subsequently regulate the pattern (Maden, 1994). Hox genes of the D cluster (former Hox 4) seem to be involved and expressed in the mesenchyme underneath the AER. When the distal part of a developing limb bud is removed, so that the domains expressing the Hox D genes are not present, the distal pattern is formed by the proximal mesenchyme after the AER makes contact. During this pattern regulation, expression of the HoxD 11 and HoxD 13 is regenerated in the proximal mesenchyme (Hayamizu et al., 1994).

These studies indicate that the AER can stimulate the regeneration of expression patterns of Hox genes in order for the complete proximal–distal pattern to be formed (Maden, 1994). In this respect, Muneoka and co-workers (1993) believe that regeneration in higher vertebrates is restricted because of the existence of regeneration barriers. According to this hypothesis, the first barrier involves wound healing (Tassava and Olsen, 1982). In higher vertebrates the AER fails to reform after amputation, leaving the stump without the necessary signals for the regeneration of the patterns. The second barrier occurs in later limb stages and possibly involves the *msx*-1 gene.

19

Regenerating versus Developing Limb

Are the mechanisms that regulate limb regeneration the same ones that regulate limb development? In other words, is regeneration a repeat of development? This is an interesting question with important implications in the control of gene expression and the mechanisms of differentiation. If, for example, the program for limb development is the same as the one for limb regeneration, then the salamander must have the ability to return very faithfully to its embryonic genetic program. This might imply that, in these animals, genes that operated during development are not silenced or that unique mechanisms of reexpression of embryonic genes might exist. If the molecular mechanisms are not the same, this implies that the program for regeneration is unique and different from the embryonic one.

Similarities in tissue development and regeneration are the subject of debate that originated virtually at the time when Spallanzani observed limb regeneration. He concluded that embryonic development and regeneration of the skeletal elements are similar. Such an idea was reiterated by observations in the late nineteenth century by Gotte, who also indicated that development and regeneration of skeletal elements are very similar in *Triton* larvae (cited by Morgan, 1901).

More recent histological work does support the notion of similarities in histogenesis between the developing and the regenerating limb tissues. Ultrastructural studies have indicated that the embryonic limb bud and the blastema are similar in the axolotl larva. Blastema cells, however, were found to possess more rough endoplasmic reticulum (Hay, 1962, 1966; Bryant et al., 1971; Faber, 1971). Differentiation of muscles in the axolotl follows almost identical patterns. The pre-muscle cells accumulate into common flexor and extensor masses and they then split (Thornton, 1938b). Skeletal differentiation was found to follow similar avenues between the two developmental phenomena (Grim and Carlson, 1974).

19.1 Patterning mechanisms

To answer the opening question, Muneoka and Bryant (1982) examined patterning mechanisms in the developing and regenerating axolotl limb. As mentioned in Chapter 16, when forelimbs are confronted with hindlimb grafts and the anterior–posterior axis is reversed, supernumerary limbs sprout from the graft site. If developing and regenerating limbs are regulated by similar patterning mechanisms, then contralateral grafting of developing limb buds to regenerating limbs (and vice versa) should also lead to the generation of supernumerary limbs, as occurred when these experiments were performed with blastema transplantation. To study this, Muneoka and Bryant (1982) employed the axolotl, where development of the hindlimb is delayed in comparison to the forelimb. At stage 46, amputation through the humerus of the forelimb produced a blastema, which at the palette stage was similar in size to the developing hindlimb bud. When these developing buds were exchanged with the blastema by ipsilateral grafting (no axis confrontation), normal development of the limbs was observed without the production of supernumerary regenerates. When contralateral grafting was performed (so that the anterior–posterior axes were confronted), supernumerary structures were produced at the graft site (Figure 19.1). These outgrowths were produced by grafting the limb bud into a regenerating limb or vice

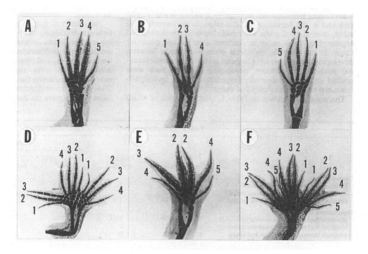

Figure 19.1 Limbs resulted from grafting involving regenerating and developing limbs in the axolotl. A: A control ipsilateral graft of a developing right hindlimb bud on a regenerating right forelimb stump. A normal 5-digit right hindlimb was formed. B: A control ipsilateral graft of a right-regenerating forelimb blastema on a developing right hindlimb bud stump. The graft produced a normal forelimb on a hindlimb stump. C: A control ipsilateral graft of a developing left hindlimb bud on a developing left hindlimb bud stump resulting in a normal 5-digit hindlimb. D: A limb resulting from a contralateral graft of a left-developing hindlimb bud to the right-regenerating forelimb blastema stump. Supernumerary limb structures are present both anterior and posterior to the grafted hindlimb. E: A limb resulting from a contralateral graft of a left-regenerating forelimb blastema on a right-developing hindlimb bud stump. Digits derived from the graft are located on the left of the figure and the supernumerary ones are on the right. The two limbs have fused during outgrowth with loss of digits. F: A limb resulting from a contralateral graft of a left-developing hindlimb bud to a right-developing hindlimb bud stump. Supernumerary limb structures are present on locations anterior and posterior to the grafted hindlimb. (Courtesy S. V. Bryant.)

versa. These experiments conclusively show that the patterning mechanisms between developing and regenerating limb are the same. These studies are important because they suggest that we can learn about limb development by studying limb regeneration, and that we can also understand the lack of limb regeneration in higher vertebrates by studying limb development.

Similarities in the patterning between the regenerating and developing limb are suggested by the need for proper epithelial–mesenchymal interactions in both. Stocum and Dearlove (1972) have transplanted limb blastema without the epithelium into the fin musculature. In one series the transplant was totally embedded. Regeneration resulted, but the regenerates were distally truncated. In other series the blastema was implanted in such a way that the distal tip protruded and could be covered by the fin wound epithelium. Smaller but distally complete regenerates were obtained. These experiments show that, as in the case of embryonic limb bud, proximal–distal organization is the result of proper epithelial–mesenchymal interactions.

19.2 The molecular evidence

If such mechanisms are the same, however, one should expect that the genes expressed in the developing and regenerating limb follow the same patterns. Molecular studies in the past few years do not agree with this idea, because expression of certain molecules is not the same in the developing and the regenerating limb. During the last decade, markers have been developed for studies during limb blastema formation. Parallel studies with the developing limb are not always feasible because the newt, for example, cannot be reared easily in the lab and reach the desired embryonic limb bud stage. Such experiments are only feasible with the axolotl or different urodeles. For example, Ferretti et al. (1989) have studied the expression of keratins in the regenerating limbs of *N. viridescens* and the developing limbs of *Pleurodeles waltl*. It was found that the keratin pair 8 and 18 was expressed in the mesenchymal blastema cells but was absent from the developing limb bud. The WE3 antibody reacts with an antigen that is up-regulated in the wound epithelium of the regenerating blastema, but is absent from the ectoderm of the developing limb bud (Tassava and Acton, 1989). Type XII collagen has been found to be up-regulated in the regenerating limb, but it is absent from the developing limb bud of *N. viridescens* and *Amblystoma*, even after amputation. Similarly, the MT5 antibody reacts with an antigen only in regenerating limbs (Tassava, 1993). Aneurogenic limbs, which are capable of regeneration, fail to express the 12/18 antigen that is readily present in the blastema of the innervated limbs. This suggests that the blastema cells can be controlled by the nerve (Fekete and Brockes, 1988). Similarly, amputation of embryonic limb buds produces a blastema that is negative for 12/18 at the early stages and positive at later ones (Fekete and Brockes, 1987). Such differences argue against the notion that regeneration is an identical copy of limb development. Instead, it seems that at least some gene expression is different between the two processes, despite the fact that patterning mechanisms seem to be the same. Reconciliation of the two ideas is a very difficult task. One obvious answer could be that the patterning mechanisms require a certain genetic program that is the same in development and regeneration, but that some regeneration mechanisms are distinct. From this point of view it is safe to say that genes responsible for patterning (that is, Hox genes) should be expressed in an identical fashion in the developing and regenerating limb (even though developing and regenerating limbs start from different places). Studies with the axolotl HoxA 9 and HoxA 13 do not support this. The expression of HoxA 13 during limb development is more distally restricted than the expression of HoxA 9. In regenerating limbs, however, both genes are expressed in the same population of blastema cells (Gardiner et al., 1995). It seems that the ability of a limb to regenerate is the result of a somewhat different program. And I say "somewhat" because many proteins specific for the developing limb are reappearing in the regenerating limb as well (Slack, 1982).

19.3 The genetic program

The fascinating events of limb regeneration and patterning raise some very interesting questions about gene regulation during limb regeneration. In other vertebrates, genes that regulate growth and differentiation and are responsible for development are controlled once the organism has reached a certain stage. Oncogenes are some of these genes. They are known to regulate growth and not to be expressed during adulthood. In fact, reexpression of such genes in the adult mammal leads to cancer formation. In adult newts, however, the developmental program seems to be always available. The only stimulus is the amputation. In this respect we should reconsider the fact that many Hox genes are expressed in the adult intact limb and that they are up-regulated after amputation (Chapter 18). Such expression is not present in the adult limbs of mammalian counterparts. The Hox genes, therefore, might provide one example of different gene control in embryonic and adult tissues in salamanders versus mammals. Despite the differences between developing and regenerating limbs, many similarities exist at the molecular level. This means that the regulation of some embryonic genes remains the same throughout the life of the animal. This in turn has amazing implications concerning gene regulation. On the other hand, if some urodeles manage to regulate some embryonic genes differently than do other vertebrates, then this is certainly unique. Needless to say, the urodeles could become experimental systems for mechanisms of gene control that could bear relevance to fundamental biological events and possibly to cancer as well.

PART III

APPLICATIONS OF
MODERN TECHNIQUES
TO THE FIELD OF LIMB
REGENERATION

20

Molecular Advances

In the past, technical problems have hindered progress in the field of urodele limb regeneration. For example, there were no cell lines for transplantations, urodeles were not easily cloned or bred (with the exception of the Mexican axolotl), transgenic animals were not available, and molecular markers and efficient vectors for gene transfer were not developed. But one has to realize that the urodeles are wonderful animals that offer us opportunities not available with other vertebrates. Most important, the urodele is capable of regenerating its limb, tail, jaw, lens, retina, and cornea. These phenomena call for much attention. After all, urodeles are the only vertebrates that have such extensive regenerative ability. We need to continue to research regeneration, but to do so we must overcome technical problems, stimulate the study of these animals, and make the urodeles once again one of the favorite animals in the field of developmental biology and genetics.

20.1 Cell culture

Cells from newt tissues can now be cultured efficiently. Since the early 1970s, cells of the newt eye could be maintained *in vitro,* so that events of dedifferentiation and transdifferentiation could be studied. Culture of blastema cells was developed in the mid-1980s and early 1990s. What is important, however, is the fact that newt limb cells have now been propagated as cell lines. This led to experiments wherein cells could be labeled (either by dyes or by reporter genes), transplanted into the amputated limb, and shown to participate and differentiate in the regenerating limb. Such experiments show the power of the *in vitro* system for studying limb regeneration. The culture system is now accessible and undeniably will contribute to the understanding of cellular mechanisms involved in limb regeneration.

20.2 Gene transfer

The success in culturing newt limb cells led to experiments involving gene transfer. We now know that these cells can be transfected as well as mammalian cells. We also learned that some idiosyncrasies might exist that are unique to the newt cells. The plasmids used in transfections seem to be in the cell episomally and do not integrate. This might represent a problem for the generation of transgenic animals, but it might be an advantage for the study of reporter gene expression; for example, the cells will keep expressing the marker upon transplantation. Indeed, such expression has been observed for up to 8 weeks after a transfected cell has been transplanted into an amputated limb (Pecorino et al., 1994). This is enough time to study the behavior of these cells during the regeneration process. Transfection of newt limb cells has been achieved with the regular methods of calcium-phosphate, with the use of lipofectamine, microinjection, and by the biolistic particle gun. This last approach is particularly interesting because of its potential to transfect cells *in vivo.* As Pecorino et al. (1994) have demonstrated, it is possible to transfect a gene of interest in the wound epithelium or into the mesenchymal blastema after removing and inverting the blastema. The transfected blastema can then be regrafted back onto the limb stump (Figure 20.1). Such an approach, along with

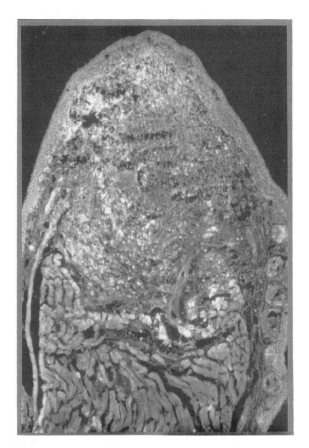

Figure 20.1 Expression of placental alkaline phosphatase in the blastema cells after *in vivo* biolistic transfection of a plasmid containing the gene for this marker. (Courtesy J. P. Brockcs.)

ideas stemming from the use of the chimeric receptors, could be valuable in the study of cell behavior during regeneration and patterning of the limb.

20.3 Vectors

If transfections such as the ones outlined in the previous section are to be performed, established vectors from other vertebrate systems, including mammalian, can be used. The newt cells respond well, and very efficiently utilize promoters from other animals. In fact, promoters from chick, human, *Xenopus,* mouse, or even those of retroviruses are capable of driving the expression of reporter genes after transfection into newt cells (Brockes, 1992; Burns et al., 1994). The inability, however, of the plasmids to integrate stably into the genome raises a serious question and leaves a puzzle to be solved. Failure of stable integration is thought to be the result of the huge genome in urodeles, but the exact mechanisms are not known. This problem must be solved if we want to study transgenic salamanders where stable integration and expression of the gene of choice are preferred. Even though newt retroviral vectors are currently available, pseudotyped retroviruses have been found to infect, integrate, and express the gene of interest after transfection into newt cells (Figure 20.2). These viruses could become the long-awaited tools for permanent gene transfer into newt cells (Burns et al., 1994).

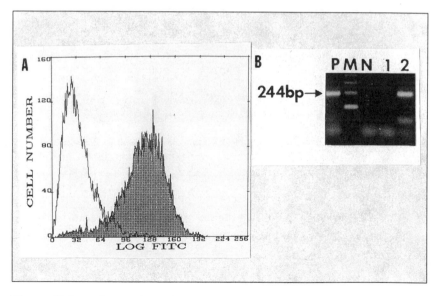

Figure 20.2 A: Expression of the reporter gene (*lacZ*) introduced into cultured limb cells via a pseudotyped retrovirus. The peak on the left represents the mock-infected cells and the peak on the right (gray) the virus-infected cells. Nearly 90 percent of the cells express the marker. B: Integration of the pseudotyped retrovirus genome into the salamander genome following infection. The expected size of a successful integration (224 bp) is seen only in the virus-infected cells (lane 2) and not in the mock-infected cells (lane 1) or the negative control (N). P is positive reaction and M is markers.

20.4 Libraries

Even though they were completely lacking just 10 years ago, genomic and cDNA libraries are now available from the most commonly used salamanders. Researchers realized that in making libraries, in principle at least, there is nothing different between urodele and mammalian libraries. Numerous genes and cDNAs have been successfully cloned from these libraries. The huge genome, however, calls for more extensive screening, which simply demands more time. Information on the existing libraries, probes, and markers is now available from the Axolotl Colony at Indiana University–Bloomington.

20.5 Transgenic animals

The use of transgenic animals has enormously facilitated studies of developmental genetics in the mouse. Similarly, the function of a particular gene can be studied by the method of homologous recombination in mice. In this procedure, the normal gene of interest is "knocked out" from embryonic stem cells and replaced by a mutated one. The technique makes use of the ability of homologous chromosomes to recombine. The cells are then transferred to a pseudopregnant mother, to produce embryos lacking the normal gene. Development can then be followed and the action of the gene analyzed in detail. The effect of the lacking gene can be seen in developmental defects in tissue where the gene product is needed. Such experiments using the axolotl or newt should first go through the stage of transgenic technology and then to homologous recombination. I believe that the generation of transgenic salamanders is only a matter of time. The availability of the pseudotyped retroviruses should make this possible. Homologous recombination, however, requires stem cells that we do not have for urodeles. In the meantime, it is already possible to generate transgenic limbs. Cells can be manipulated *in vitro,* selected, and transplanted into an amputated limb. The cells will take part in regeneration and become permanent residents of the regenerated limb. Amputation of such limbs will, in theory, produce a regenerate that will be marked genetically. In principle, the regenerating limbs will be transgenic. With the development of this procedure, we can then bypass the need for transgenic animals, and this, in my opinion, can open new avenues in the study of limb regeneration.

Advances in molecular techniques with the newt and axolotl systems as outlined here have already told us that every experiment feasible in mammals should be feasible with the newt or axolotl as well. And, despite some early problems dealing with the uniqueness of the urodele system, these techniques will undoubtedly contribute to the field of regeneration, expand it, and meet the expectations of the scientific community as we enter the twenty-first century.

EPILOGUE

The Marriage of Solitude and Enthusiasm

The salamanders must go on. . .

—Takis Alexiou

The phenomenon of limb regeneration has been known for more than 200 years. As mentioned in the prologue, regeneration of limbs was first reported by Spallanzani in 1768. As we enter the fourth century after its discovery, we cannot stop thinking of the ups and downs of the field. Many scientists, myself often included, believe that the field of limb regeneration has been neglected by the scientific community– and in some cases even abandoned by its students. The difficulty in working with the system was understandable in the past, but not any longer. The field of limb regeneration always had very devoted students who committed their entire lives to understanding it. The fact that we still have few clues about the genetics of the system should not be a reason to abandon it. And even if it has been somewhat in solitude while other systems have advanced, we–the people who enthusiastically study it–believe that there is no better time than now to feel optimistic about what is coming. The field of limb regeneration will prove to be one of the most important in the biomedical world. Its relevance to cell differentiation, induction, and pattern formation could provide clues that other systems cannot. Ayn Rand, in her epic novel *Atlas Shrugged,* tells how Atlas, already carrying the world on his shoulders, is presented with an additional challenge above his duties. His reply was simply to shrug and prevail.

References

Acampora, D., D'Esposito, M., Faiella, A., Pannese, M., Migliaccio, E., Morelli, F., Stornaiuolo, A., Nigro, V., Simeone, A., and Boncinelli, E. (1990). The human HOX gene family. *Nucl. Acids Res. 17*: 10385–402.

Ahrens, P. B., Solursh, M., and Reiter, R. S. (1977). Stage-related capacity for limb chondrogenesis in cell culture. *Dev. Biol. 60*: 69–82.

Anton, H. J. (1965). The origin of blastema cells and protein synthesis during forelimb regeneration in *Triturus*. In *Proceedings, Regeneration in Animals,* Eds. V. Kiortsis and H.A.L. Trampusch, pp. 377–95. Amsterdam: North-Holland.

Balls, M., and Ruben, L. N. (1964). Variation in the response of *Xenopus laevis* to normal tissue homografts. *Dev. Biol. 10:* 92–104.

Bantle, J. A., and Tassava, R. A. (1974). The neurotrophic influence on RNA precursor incorporation into polyribosomes of regenerating adult newt forelimbs. *J. Exp. Zool. 189:* 101–14.

Bao, C., Singer, M., and Ilan, J. (1986). Effects of forelimb amputation and denervation on protein synthesis in spinal cord ganglia of the newt. *Proc. Natl. Acad. Sci. USA 83:* 7971–74.

Barber, L. W. (1944). Correlations between wound healing and regeneration in forelimbs and tails of lizards. *Anat. Rec. 89:* 441–53.

Bast, R. E., Singer, M., and Ilan, J. (1979). Nerve-dependent changes in content of ribosomes, polysomes and nascent peptides in newt limb regenerates. *Dev. Biol. 70:* 13–26.

Bateson, W. (1894). *Materials for the Study of Variation.* London: MacMillan.

Bazzoli, A. S., Wanson, J., Scott, W. J., and Wilson, J. G. (1977). The effects of thalidomide and two analogs on the regenerating forelimb of the newt. *J. Embryol. Exp. Morphol. 41:* 125–35.

Beauchemin, M., and Savard, P. (1991). Graded expression of two homeo box genes in the adult newt limb and their corresponding blastema. Presented at the Midwest Regional Developmental Biology Conference, West Lafayette, IN, 1991.

Beauchemin, M., and Savard, P. (1992). Two *distal-less* related homeobox-containing genes expressed in regeneration blastemas of the newt. *Dev. Biol. 154:* 55–65.

Beauchemin, M., and Savard, P. (1993). Expression of five homeobox genes in the adult newt appendages and regeneration blastemas. In *Limb Development and Regeneration,* eds. J. F. Fallon et al., pp. 41–50. New York: Wiley-Liss.

Beauchemin, M., Noiseux, N., Tremblay, M., and Savard, P. (1994). Expression of Hox A11 in the limb and the regeneration blastema of adult newt. *Int. J. Dev. Biol. 38:* 641–49.

Beauchemin, M., Noiseux, N., Tremblay, M., and Savard, P. (1995). Expression of Hox A3 in the limb and the regeneration blastema of adult newt. *Gene* (submitted).

Beauchemin, M., Tremblay, M., and Savard, P. (1995). Graded expression of an *ems*-like homeobox gene (NvHBox-6) in the adult newt limb and its corresponding blastema. *Development* (submitted).

Becker, R. O. (1972). Stimulation of partial limb regeneration in rats. *Nature 235:* 109–11.

Bellairs, A., d'A., and Bryant, S. V. (1968). Effects of amputation of limbs and digits of lacertid lizards. *Anat. Rec. 161:* 489–96.

Belleville, S., Beauchemin, M., Tremblay, M., Noiseux, N., and Savard, P. (1992). Homeobox-containing genes in the newt are organized in clusters similar to other vertebrates. *Gene 114:* 179–86.

Benjamin, R. C. and Gill, D. M. (1980). Poly(ADP-ribose) synthetase *in vitro* programmed by damaged DNA. *J. Biol. Chem. 255:* 10502–508.

Bischler, V. (1926). L'influence du squelette dans la régéneration et le potentialités des divers territoires du membres chez les *Triton cristatus. Rev. Suisse Zool. 33:* 431–560.

Bodemer, C. W. (1958). The development of nerve-induced supernumerary limbs in the adult newt, *Triturus viridescens. J. Morphol. 102:* 555–82.

Bodemer, C. W., and Everett, N. B. (1959). Localization of newly synthesized proteins in regenerating newt limbs as determined by radioautographic localization of injected methionine-S[35]. *Dev. Biol. 1:* 327–42.

Boilly, B., and Albert, P. (1990). *In vitro* control of blastema cell proliferation by extracts from epidermal cap and mesenchyme of regenerating limbs of axolotls. *Roux's Arch. Dev. Biol. 198:* 443–47.

Boilly, B., Cavanaugh, K. P., Thomas, D., Hondermarck, H., Bryant, S. V., and Bradshaw, R. A. (1991). Acidic fibroblast growth factor is present in regenerating limb blastemas of axolotls and binds specifically to blastema tissues. *Dev. Biol. 145:* 302–10.

Borgens, R. B., Vanable, J. W., Jr., and Jaffe, L. F. (1977a). Bioelectricity and Regeneration. *J. Exp. Zool. 200:* 403–16.

Borgens, R. B., Vanable, J. W., Jr., and Jaffe, L. F. (1977b). Bioelectricity and regeneration: Large currents leave the stumps of regenerating newt limbs. *Proc. Natl. Acad. Sci. USA 74:* 4528–32.

Borgens, R. B., Vanable, J. W., Jr., and Jaffe, L. F. (1979a). Reduction of sodium dependent stump currents disturbs urodele limb regeneration. *J. Exp. Zool.* 209: 377–86.

Borgens, R. B., Vanable, J. W., Jr., and Jaffe, L. F. (1979b). Small artificial currents enhance *Xenopus* limb regeneration. *J. Exp. Zool. 207:* 217–26.

Borgens, R. B. (1982). Mice regrow tips of their foretoes. *Science 217:* 747–50.

Borgens, R. B., McGinnis, M. E., Vanable, J. W., Jr., and Miles, E. S. (1984). Stump currents in regenerating salamanders and newts. *J. Exp. Zool. 231:* 249–56.

Breedis, C. (1952). Induction of accessory limbs and of sarcoma in the newt (*Triturus viridescens*) with carcinogenic substances. *Cancer Res. 12:* 861–73.

Brockes, J. P. (1984). Mitogenic growth factors and nerve dependence of limb regeneration. *Science 225:* 1280–7.

Brockes, J. P., and Kintner, C. R. (1986). Glial growth factor and nerve-dependent proliferation in the regeneration blastema of urodele amphibians. *Cell 45:* 301–6.

Brockes, J. P. (1989). Retinoids, homeobox genes, and limb morphogenesis. *Neuron 2:* 1285–94.

Brockes, J. P. (1992). Introduction of a retinoid reporter gene into the urodele limb blastema. *Proc. Natl. Acad. Sci. USA 89:* 11386–90.

Brown, R., and Brockes, J. P. (1991). Identification and expression of a regeneration-specific homeobox gene in the newt limb blastema. *Development 111:* 489–96.

Brunst, V. V. (1950). Influence of X-rays on limb regeneration in urodele amphibians. *Quart. Rev. Biol. 25:* 1–29.

Brunst, V. V. (1961). Some problems of regeneration. *Quart. Rev. Biol. 36:* 178–206.

Bryant, P. J., Bryant, S. V., and French, V. (1977). Biological regeneration and pattern formation. *Sci. Am. 237:* 66–81.

Bryant, S. V., Fyfe, D., and Singer, M. (1971). The effects of denervation on the ultrastructure of young limb regenerates in the newt *Triturus*. *Dev. Biol. 24:* 577–95.

Bryant, S. V. (1976). Regenerative failure of double half limbs in *Notophthalmus viridescens*. *Nature 263:* 676–79.

Bryant, S. V., and Iten, L. E. (1976). Supernumerary limbs in amphibia: Experimental production in *Notophthalmus viridescens* and a new interpretation of their formation. *Dev. Biol. 50:* 212–34.

Bryant, S. V., and Baca, B. A. (1978). Regenerative ability of double-half and half upper arms in the newt, *Notophthalmus viridescens*. *J. Exp. Zool. 204:* 307–24.

Bryant, S. V., French, V., and Bryant, P. J. (1981). Distal regeneration and symmetry. *Science 212:* 993–1002.

Bryant, S. V., and Gardiner, D. M. (1992). Retinoic acid, local cell–cell interactions and pattern formation in vertebrate limbs. *Dev. Biol. 152:* 1–25.

Bryant, S. V. (1994). Limb development and regeneration: cell–cell interactions, retinoic acid and Hox genes. Presented at the 34th Midwest Regional Developmental Biology Conference, Granville, OH, May 23–25.

Burnet, B. M., and Liversage, R. A. (1964). The influence of chloramphenicol on the regenerating forelimb in adult *Triturus viridescens*. *Am. Zool. 4:* 427.

Burns, J. C., Friedmann, T., Driever, W., Burrascano, M., and Yee, L.-K. (1993). Vesicular stomititis virus G glycoprotein pseudotyped retroviral vectors: Concentration to very high titer and efficient gene transfer into mammalian and nonmammalian cells. *Proc. Natl. Acad. Sci. USA 90:* 8033–37.

Burns, J. C., Matsubara, T., Lozinski, G., Yee, J.-K., Friedmann, T., Washabaugh, C.H., and Tsonis, P. A. (1994). Pantropic retroviral vector-mediated gene transfer, integration, and expression in newt limb blastema cells. *Dev. Biol. 165:* 285–289.

Butler, E. G. (1933). The effects of X-irradiation on the regeneration of the forelimbs of *Amblystoma* larvae. *J. Exp. Zool. 65:* 271–315.

Butler, E. G., and O'Brien, J. P. (1942). Effects of localized X-irradiation on regeneration of the urodele limb. *Anat. Rec. 84:* 407–13.

Butler, E. G., and Schotte, O. E. (1949). Effects of delayed denervation on regenerative activity in limbs of urodele larvae. *J. Exp. Zool. 112:* 361–92.

Butler, E. G., and Blum, H.F. (1955). Regenerative growth in the urodele forelimb following ultraviolet radiation. *J. Natl. Cancer Inst. 15:* 877–87.

Butler, E. G. (1955). Regeneration of the urodele forelimb after reversal of its proximal–distal axis. *J. Morphol. 96:* 265–82.

Butler, E. G., and Blum, H. F. (1963). Supernumerary limbs of urodele larvae resulting from localized ultraviolet light. *Dev. Biol. 7:* 218–33.

Cameron, S. A., Hilgers, A. R., and Hintenberger, T. J. (1986). Evidence that reserve cells are a source of regenerated adult newt muscle *in vitro*. *Nature 310:* 607–8.

Caplan, A. I. (1991). Mesenchymal stem cells. *J. Orthop. Res.* 9: 641–50.

Cardoso, M. C., Leonhardt, H., and Nadal-Ginard, B. (1993). Reversal of terminal differentiation and control of DNA replication: Cyclin A and cdk2 specifically localize at sunbuclear sites of DNA replication. *Cell* 74: 979–92.

Carlone, R. L, and Foret, J. E. (1979). Stimulation of mitosis in cultured limb blastemata of the newt *Notophthalmus viridescens*. *J. Exp. Zool. 210:* 245–52.

Carlone, R. L., and Boulianne, R. P. (1991). Identification of proteins potentially involved in proximal–distal pattern formation in the regenerating forelimb of the newt. *Biochem. Cell Biol. 70:* 285–90.

Carlson, B. M. (1967a). The effect of actinomycin D upon epidernal–mesodermal interactions in limb regeneration. *Am. Zool. 7:* 702.

Carlson, B. M. (1967b). The histology of inhibition of limb regeneration in the newt *Triturus* by actinomycin D. *J. Morphol. 122:* 249–64.

Carlson, B. M. (1969). Inhibition of limb regeneration in the axolotl after treatment of the skin with actinomycin D. *Anat. Rec. 163:* 389–402.

Carlson, B. M. (1971). The distribution of supernumerary limb-inducing capacity in tissues of *Rana pipiens*. *Oncology 25:* 365–71.

Carlson, B. M. (1972). Muscle morphogenesis in axolotl limb regenerates after removal of stump musculature. *Dev. Biol. 28:* 487–97.

Carlson. B. M. (1974a). Factors controlling the initiation and cessation of early events in the regenerative process. In *Neoplasia and Cell Differentiation,* ed. G. V. Sherbet, pp. 60–105. Basel: Karger.

Carlson, B. M. (1974b). Morphogenetic interactions between rotated skin cuffs and underlying stump tissues in regenerating axolotl forelimbs. *Dev. Biol. 39:* 263–85.

Carlson, B. M. (1975a). The effects of rotation and positional change of stump tissues upon morphogenesis of the regenerating axolotl limb. *Dev. Biol. 47:* 269–91.

Carlson, B. M. (1975b). Multiple regeneration from axolotl limb stumps bearing cross-transplanted minced muscle regenerate. *Dev. Biol. 45:* 203–8.

Carlson, B. M. (1979). Relationship between tissue and epimorphic regeneration of skeletal muscle. In *Muscle Regeneration*, pp. 57–71. New York: Raven Press.

Carroll, J. M., and Sicard, R. E. (1980). Cyclic nucleotide metabolism during amphibian forelimb regeneration. I. The cyclic AMP phosphodiesterases. *Roux's Arch. Dev. Biol. 189:* 107–10.

Casimir, C. M., Gates, P. B., Ross-Macdonald, P. B., Jackson, J. F., Patient, R. K., and Brockes, J. P. (1988a). Structure and expression of a newt cardioskeletal myosin gene. *J. Mol. Biol. 202:* 287–96.

Casimir, C. M., Gates, P. B., Patient, R. K., and Brockes, J. P. (1988b). Evidence for dedifferentiation and metaplasia in amphibian limb regeneration from inheritance of DNA methylation. *Development 104:* 657–68.

Chalkley, D. T. (1954). A quantitative histological analysis of forelimb regeneration in *Triturus viridescens. J. Morphol. 94:* 21–70.

Chalkley, D. T. (1956). The cellular basis of limb regeneration. In *Regeneration in Vertebrates*, ed. C. S. Thornton, pp. 34–58. University of Chicago Press.

Chan, W. Y., Lee, K. K. H., and Tam, P. P. L. (1991). Regenerative capacity of forelimb buds after amputation in mouse embryos at the early-organogenesis stage. *J. Exp. Zool. 260:* 74–83.

Cherkasova, L. V. (1974). Autoradiographic investigation in the process of experimental restoration of lost regenerative capacity of the limbs of the tailess amphibia. *DAN SSSR 215:* 1505–08.

Cherkasova, L. V. (1982). Postsatellites in muscular tissue in adult tailed amphibia. *DAN SSSR 267:* 1235–36.

Choo, A. F., Logan, D. M., and Rathbone M. P. (1978). Nerve trophic effects: An *in vitro* assay for factors involved in regulation of protein synthesis in regenerating amphibian limbs. *J. Exp. Zool. 206:* 347–54.

Cohen, S. M., Bronner, G., Kuttner, F., Jurgens, G., and Jackle, H. (1989). *Distal-less* encodes a homoeodomain protein required for limb development in *Drosophila. Nature 338:* 432–34.

Cohn, M. J., Izpisúa-Belmonte, J. C., Abud, H., Heath, J. K., and Tickle, C. (1995). Fibroblast growth factors induce additional limb development from the flank of chick embryos. *Cell 80:* 739–46.

Costaridis, P., Zafeiratos, C., Kiortis, V., and Papageorgiou, S. (1989). Diverse supernumerary structures develop after inverting the anteroposterior limb axis of the anuran. *Dev. Biol. 132:* 502–11.

Costaridis, P., Papageorgiou, S., Kiortsis, V., and Zafeiratos, C. (1991). Types of supernumerary outgrowths produced after inverting the dorsoventral limb axis of the anuran *Bufo bufo. Roux's Arch. Dev. Biol. 200:* 104–7.

Crawford, K., and Stocum, D.L. (1988). Retinoic acid coordinately proximal-izes regenerate pattern and blastema differential affinity in axolotl limbs. *Development 102:* 687–98.

David, L. (1932). Das Verhalten von Extremitäten regeneraten des weißen und pigmentierten Axolotl bei heteroplastischer, heterotopen und ortho-topen Transplantation und sukzeissiver Regeneration. *Roux's Arch. Entwicklungsmech. Org.* 126: 457–511.

Davis, A. P., and Capecchi, M. R. (1994). Axial homeosis and appendicular skeleton defects in mice with a targeted disruption of Hox D11. *Development 120:* 2187–98.

Davis, A. P., Witte, D. P., Hsieh-Li, H. M., Potter, S. S., and Capecchi, M.R. (1995). Absence of radius and ulna in mice lacking Hox A11 and Hox D11. *Nature 375:* 791–5.

Dearlove, G. E., and Stocum, D. L. (1974). Denervation-induced changes in soluble protein synthesis content during forelimb regeneration in the adult newt *Notophthalmus viridescens. J. Exp. Zool. 190:* 317–27.

Dearlove, G. E., and Dresden, M. H. (1976). Regenerative abnormalities in *Notophthalmus viridescens* induced by repeated amputations. *J. Exp. Zool. 196:* 251–62.

De Both, N. J. (1970). The developmental potencies of the regeneration blastema of the axolotl limb. *Roux's Arch. Entwicklungsmech. Org. 165:* 242–76.

Deck, J. D., and Riley, H. W. (1958). Regenerates on hindlimbs with reversed proximal–distal polarity in larval and metamorphosing urodeles. *J. Exp. Zool. 138:* 493–504.

Deck, J. D., and Dent, J. N. (1970). The effects of infused material upon regeneration of newt limbs. III. Blastemal extracts and alkaline phos-phatase in irradiated limb stumps. *Anat. Rec. 168:* 525–35.

Della Valle, P. (1913). La doppia rigenerazione inversa nella Frature della zampe di Triton. *Boll. Soc. Natur. Napoli 25:* 73–85.

Del Rio-Tsonis, K., and Tsonis, P. A. (1994). Oncogene-related sequences in amphibian genomes. *Oncol. Rep. 1:* 847–48.

Del Rio-Tsonis, K., Washabaugh, C. H., and Tsonis, P. A. (1992). The mutant axolotl Short toes exhibits impaired limb regeneration and abnormal basement membrane formation. *Proc. Natl. Acad. Sci. USA 89:* 5502–6.

Dent, J. M. (1962). Limb regeneration in larvae and metamorphosing individ-uals of South African clawed toad. *J. Morphol. 110:* 61–78.

Desselle, J. C., and Gontcharof, M. (1978). Cytophotometric detection of the participation of cartilage grafts in regeneration of X-rayed urodele limbs. *Biol. Cell 33:* 45–53.

Deuchar, E. M. (1976). Regeneration of amputated limb buds in early rat embryos. *J. Embryol. Exp. Morphol. 35:* 345–54.

Di Giorgi, P. (1924). Les potentialités des régénérats chez *Salamandra macu-losa.* Croissance et differenciation. *Rev. Suisse Zool. 31:* 1–53.

Dinsmore, C. E. (1974). Morphogenetic interactions between minced muscle and transplanted blastemas in the axolotl. *J. Exp. Zool. 187:* 223–32.

Dinsmore, C. E. (1991). *A History of Regeneration Research: Milestones in the Evolution of a Science.* Cambridge University Press.

Dolle, P., Ruberte, E., Kastner, P., Petkovich, M., Stoner, C. M., Gudas, L. J., and Chambon, P. (1989a). Differential expression of genes encoding alpha, beta and gamma retinoic acid receptors and CRABP in the developing limbs of the mouse. *Nature 342:* 702–5.

Dolle, P., Izpisúa-Belmonte, J-C., Flakestein, H., Renucci, A., and Duboule, D. (1989b). Coordinate expression of the murine HOX-5 complex homoeo-box-containing genes during limb pattern formation. *Nature 342:* 767–72.

Dresden, M. H., and Gross, J. (1970). The collagenolytic enzyme in the regenerating limbs of the newt *Triturus viridescens. Dev. Biol. 22:* 129–37.

Driesch, H. (1902). Studien uber das regulations vermogen der organismen. Die restitution der *Clavellina lepadiformis. Roux's Arch. Entwick lungsmech. 14:* 247.

Duboule, D. (1991). Patterning in the vertebrate limb. *Curr. Opin.Genetics and Development 1:* 211–16.

Dunis, D., and Namenwirth, M. (1977). The role of grafted skin in the regeneration of X-irradiated axolotl limbs. *Dev. Biol. 56:* 97–109.

Eagleson, G. W. (1993). Hyperproduction and secretion of prolactin-like hormones (PRL) in the eyeless (e/e) mutant axolotl: Effects upon regeneration. Presented at the International Workshop on the Molecular Biology of Axolotls and Other Urodeles, Indianapolis, IN, October 13–16.

Edelman, G. M. (1988). *Topobiology.* New York: Basic Books.

Eguchi, G. (1988). Cellular and molecular background of Wolffian lens regeneration. In *Regulatory Mechanisms in Developmental Processes,* eds. G. Eguchi, T. S. Okada, and L. Saxen, pp. 147–58. New York: Elsevier.

Eguchi, G., and Watanabe, K. (1973). Elicitation of lens formation from the ventral iris epithelium of the newt by a carcinogen, N-methyl-N'-nitro-N-nitrosoguanidine. *J. Embryol. Exp. Morphol. 30:* 63–71.

Emerson, H. S. (1940). Embryonic induction in regenerating tissue of *Rana pipiens* and *Rana clamitans* larvae. *J. Exp. Zool. 83:* 191–222.

Engel, A. G., and Biesecker, G. (1982). Complement activation in muscle fiber necrosis: Demonstration of the membrane attack complex of compliment in necrotic fibers. *Ann. Neurol. 12:* 289–96.

Estrada, C. M., Park, C. D., Castilla, M., and Tassava, R. A. (1993). Monoclonal antibody WE6 identifies an antigen that is upregulated in the wound epithelium of newts and frogs. In *Limb Development and Regeneration,* eds. J. F. Fallon et al., pp. 271–82. New York: Wiley-Liss.

Evans, R. M. (1988). The steroid and thyroid hormone receptor superfamily. *Science 240:* 889–95.

Faber, J. (1960). An experimental analysis of regional organization in the regenerating forelimb of the axolotl (*Ambystoma mexicanum*). *Arch. Biol. 71:* 1–67.

Faber, J. (1962). Additional experiments on the self-differentiation of transplanted whole and half forelimb regenerates of *Amblystoma mexicanum. Arch. Biol.73:* 269–378.

Faber, J. (1971). Vertebrate limb ontogeny and limb regeneration: morphogenetic parallels. *Adv. Morphog. 9:* 127–47.

Fallon, J. F., Lopez, A., Ros, M. A., Savage, M. P., Olwin, B. B., and Simandl, B. K. (1994). FGF 2: Apical ectodermal ridge growth signal for chick limb development. *Science 264:* 104–7.

Farinella-Ferruzza, N. (1950). Risultati di esperimenti di trapianti eterogenei dell'abbozzo codale negli amfibi. *Boll. Zool. 17:* 113–9.

Farinella-Ferruzza, N. (1953). Risultati di trapianti di bottone codale di urodeli su anuri e vice versa. *Rivista di Biol. 45:* 523–57.

Farinella-Ferruzza, N. (1956). The transformation of a tail into a limb after xenoplastic transplantation. *Experientia 15:* 304–5.

Farzaneh, F., Zalin, R., Brill, D., and Shall, S. (1982). DNA strand breaks and ADP-ribosyl transferase activation during cell differentiation. *Nature 300:* 362–6.

Fedotov, D. M. (1946). Russian work on chemical induction in adult animals. *Nature 159:* 367–8.

Fekete, D. M., and Brockes, J. P. (1987). A monoclonal antibody detects a difference in the cellular composition of developing and regenerating limbs of the newts. *Development 99:* 589–602.

Fekete, D. M., and Brockes, J. P. (1988). Evidence that the nerve controls molecular identity of progenitor cells for limb regeneration. *Development 103:* 567–73.

Ferretti, P., and Brockes, J. P. (1988). Culture of newt cells from different tissues and their expression of a regeneration-associated antigen. *J. Exp. Zool. 247:* 77–91.

Ferretti, P., Fekete, D. M., Patterson, M., and Lane, E. B. (1989). Transient expression of simple epithelial keratins by mesenchymal cells of regenerating newt limb. *Dev. Biol. 133:* 415–24.

Ferretti, P., and Brockes, J. P. (1990). The monoclonal antibody 22/18 recognizes a conformational change in an intermediate filament of the newt, *Notophthalmus viridescens*, during limb regeneration. *Cell Tissue Res. 259:* 483–93.

Ferretti, P., Brockes, J. P,. and Brown, R. (1991). A newt type II keratin restricted to normal and regenerating limbs and tails is responsive to retinoic acid. *Development 111:* 497–507.

Ferretti, P., Corcoran, J. P. T., and Ghosh, S. (1993). Expression and regulation of keratins in the wound epithelium and mesenchyme of the regenerating newt limb. In *Limb Development and Regeneration,* eds. J. F. Fallon et al., pp. 261–9. New York: Wiley-Liss.

Fimian, W. J. (1959). The *in vitro* cultivation of amphibian blastema tissue. *J. Exp. Zool 140:* 125–43.

Fleming, M. W., and Tassava, R. A. (1981). Preamputation and postamputation histology of the neonatal opossum hindlimb: Implications for regeneration experiments. *J. Exp. Zool. 215:* 143–9.

Fowler, I., and Sisken, B. F. (1982). Effects of augmentation of nerve supply upon limb regeneration in the chick embryo. *J. Exp. Zool. 221:* 49–59.

Francoeur, R. T. (1968). General and selective inhibition of amphibian regeneration by vinblastine and actinomycin. *Oncology 22:* 218–26.

Fritsch, C. (1911). Experimentelle Studien über Regenerationsvorganger des Gleidmassenskeletts. *Zool. Jahrb Abt. Physiol. 30:* 377–472.

Fuentes, E. J., Mescher, A. L., Ekman, R., and Tsonis, P. A. (1993). Expression of hydra head activator in newt tissues and effects on limb regeneration. *In Vivo 7:* 59–64.

Gardiner, D. M., and Bryant, S. V. (1989). Organization of positional information in the axolotl limb. *J. Exp. Zool. 251:* 47–55.

Gardiner, D.M., Blumberg, B. and Bryant, S.V. (1993). Expression of homeobox genes in limb regeneration. In *Limb Development and Regeneration,* eds. J. F. Fallon et al., pp. 31–40. New York: Wiley-Liss.

Gardiner, D. M., Komine, Y., Mullen, L., and Bryant, S. V. (1993). Molecular approaches to limb regeneration and development in axolotls. Presented at the 12th Annual Marcus Singer Symposium, Irvine, CA, December 3–4.

Gardiner, D. M., Blumberg, B., Komine, Y., and Bryant, S. V. (1995). Regulation of HoxA expression in developing and regenerating axolotl limbs. *Development 121:* 1731–41.

Garling, D. J., and Tassava, R. A. (1984). Injury, nerve, and wound epidermis related electrophoretic and fluorographic protein patterns in forelimbs of adult newts. *J. Exp. Zool. 231:* 221–41.

Gebhardt, D. O. E., and Faber, J. (1966). The influence of aminopterin on limb regeneration in *Amblystoma mexicanum. J. Embryol. Exp. Morphol. 16:* 143–58.

Geduspan, J. S., and Solursh, M. (1992). A growth-promoting influence from mesonephros during limb outgrowth. *Dev. Biol. 151:* 242–50.

Gehring, W. J. (1985). The molecular basis of development. *Sci. Am 253:* 153–62.

Geraudie, J., and Singer, M. (1978). Nerve dependent macromolecular synthesis in the epidermis and blastema of the adult newt regenerate. *J. Exp. Zool. 203:* 455–60.

Geraudie, J., Hourdry, J., Vriz, S., Singer, M., and Mechali, M. (1990). Enhanced *c-myc* gene expression during forelimb regenerative outgrowth in the young *Xenopus laevis. Proc. Natl. Acad. Sci. USA 87:* 3797–801.

Geraudie, J., Monnot, M. J., Ridet, A., Thorogood, P., and Ferretti, P. (1993). Is exogenous retinoic acid necessary to alter positional information during regeneration of the fin in zebrafish? In *Limb Development and Regeneration,* eds. J. F. Fallon et al., pp. 803–14. New York: Wiley-Liss.

Gibbins, J. R. (1978). Epithelial migration in organ culture. A morphological and time lapse cinematographic analysis of migrating squamous epithelium. *Pathology 10:* 207–18.

Giguere, V., Ong, E. S., Segui, P., and Evans, R. M. (1987). Identification of a receptor for the morphogen retinoic acid. *Nature 330:* 624–9.

Giguere, V., Ong, E. S., Evans, R. M., and Tabin, C. J. (1989). Spatial and temporal expression of the retinoic acid receptor in the regenerating amphibian limb. *Nature 337:* 566–9.

Giguere, V., Lyn, S., Yip, P., Siu, C. H., and Amin, S. (1990). Molecular cloning of cDNA encoding a second cellular retinoic acid-binding protein. *Proc. Natl. Acad. Sci. USA 87:* 6233–7.

Giguere, V. (1994). Retinoic acid receptors and cellular retinoid binding proteins: Complex interplay in retinoid signalling. *Endocrine Rev. 15:* 61–79.

Glass, L. (1977). Patterns of supernumerary limb regeneration. *Science 198:* 321–2.

Globus, M. (1978). Neurotrophic contribution to a proposed tripartite control of the mitotic cycle in the regeneration blastema of the newt *Notophthalmus (Triturus) viridescens. Am. Zool. 18:* 855–68.

Globus, M., and Vethamany-Globus, S. (1985). *In vitro* studies of controlling factors in newt limb regeneration. In *Regulation of Vertebrate Limb Regeneration,* ed. R. E. Sicard, pp. 106–27. New York: Oxford University Press.

Globus, M., Vethamany-Globus, S., and Kesik, A. (1987). Control of blastema cell proliferation by possible interplay of calcium and cyclic nucleotides during newt limb regeneration. *Differentiation 35:* 94–9.

Globus, M., and Alles, P. (1990). A search for immunoreactive substance P and other neural peptides in the limb regenerate of the newt *Notophthalmus viridescens. J. Exp. Zool. 254:* 165–76.

Godlewski, K. (1904). Der einfluss des zentralnervensystems auf die regeneration bei Tritonen. *C. R. Cong. Int. Zool. Berne* (cited by Singer, 1952).

Godovac-Zimmerman, J. (1988). The structural motif of ß-lactoglobulin and retinol-binding protein: A basic framework for binding and transport of small hydrophobic molecules? *Trends Biochem. Sci. 13:* 64–6.

Goldhamer, D.J., Tomlinson, B. J., and Tassava, R. A., (1989). Developmentally regulated wound epithelial antigen of the newt limb regenerate is also present in a variety of secretory/transport cell types. *Dev. Biol. 135:* 392–404.

Goldhamer, D. J., Tomlinson, B. L., and Tassava, R. A. (1992). Ganglia implantation as a means of supplying neurotrophic stimulation to the newt regeneration blastema: Cell-cycle effects in innervated and denervated limbs. *J. Exp. Zool. 262:* 71–80.

Gomes, L. (1964). Effects of N-dichloroacetyl-dl-serine in regeneration in the forelimb of *Diemictilus viridescens. Proc. Soc. Exp. Biol. Med. 115:* 204–6.

Gospodarowicz, D. (1974). Localization of a fibroblast growth factor and its effects alone and with hydricortisone on 3T3 cell growth. *Nature 249:* 123–7.

Gospodarowicz, D., and Mescher, A. L. (1981). Fibroblast growth factor and vertebrate regeneration. In *Neurofibromatosis (Von Recklingshausen Disease),* eds. V. M. Riccardi and J. J. Malvihill, pp. 149–71. New York: Raven Press.

Goss, R. J. (1956a). Regenerative inhibition following amputation and immediate insertion into the body cavity. *Anat. Rec. 126:* 15–27.

Goss, R. J. (1956b). The relation of bone to the histogenesis of cartilage in regenerating forelimbs and tails of adult *Triturus viridescens. J. Morphol. 98:* 89–123.

Goss, R. J. (1969). *Principles of Regeneration.* New York: Academic Press.

Goss, R. J. (1992). The evolution of regeneration: Adaptive or inherent? *J. Theor. Biol. 159:* 241–60.

Goss, R. J., and Holt, R. (1992). Epimorphic vs. tissue regeneration in Xenopus forelimbs. *J. Exp. Zool. 261:* 451–7.

Greenberg, G., and Hay, E. D. (1982). Epithelia suspended in collagen gels can lose polarity and express characteristics of migrating mesenchymal cells. *J. Cell Biol. 95:* 333.

Griffin, K. J. P., Fekete, D. M., and Carlson, B. M. (1987). A monoclonal antibody stains myogenic cells in regenerating newt muscle. *Development 101:* 267–77.

Grillo, H. C., Lapiere, C. M., Dresden, M. H., and Gross, J. (1968). Collagenolytic activity in regenerating forelimbs of the adult newt (*Triturus viridescens*). *Dev. Biol. 17:* 571–83.

Grim, M., and Carlson, B. M. (1974). A comparison of morphogenesis of muscles of the forearm and hand during ontogenesis and regeneration in the axolotl (*Amblystoma mexicanum*). II. The development of musclular patterns in the embryonic and regenerating limb. *Z. Anat. Entwickl. Gesch. 145:* 149–67.

Groell, A. L., Gardiner, D. M., and Bryant, S. V. (1993). Stability of positional identity of axolotl blastema cells *in vitro. Roux's Arch. Dev.. Biol. 202:* 170–5.

Gulati, A. K., Zalewski, A. A. and Reddi, A. H. (1983). An immunofluorescent study of the distribution of fibronectin and laminin during limb regeneration in the adult newt. *Dev. Biol. 96:* 355–65.

Gurdon, J. B., Harger, P., Mithchell, A., and Lemaire, P. (1994). Activin signalling and response to a morphogen gradient. *Nature 371:* 487–92.

Guyenot, E., and Shotte, O. E. (1926). Le rôle du système nerveux dans l'édification des régénérats de pattes chez les urodeles. *C. R. Soc. Biol. Paris 94:* 1050–2.

Guyenot, E., and Schotte, O. E. (1927). Greffe de régénérat et différenciation induite. *C. R. Soc. de Phys. Hist. Nat. Genève 44:* 21–3.

Guyenot, E. (1927). Le problème morphogénétique dans la régénération des urodeles: détermination et potentialités des régénérats. *Rev. Suisse Zool. 34:* 127–55.

Guyenot, E., Dinichert-Favarger, J., and Galland, M. (1948). L'exploration du territoire de la patte antérieure du *Triton. Rev. Suisse Zool. 55:* 1–120.

Hall, A. B., and Schotte, O. E. (1951). Effects of hypophysectomies upon the initiation of regenerative processes in the limb of *Triturus viridescens. J. Exp. Zool. 118:* 363–88.

Harrison, R. G. (1921). On relations of symmetry in transplanted limbs. *J. Exp. Zool. 32:* 1–136.

Hay, E. D. (1956). Effects of thyroxine on limb regeneration in the newt *Triturus viridescens. Bull. John Hopkins Hosp. 99:* 262–85.

Hay, E. D., and Fischman, D. A. (1961). Origin of the blastema in regenerating limbs of the newt *Triturus viridescens. Dev. Biol. 3:* 26–59.

Hay, E. D. (1962). Cytological studies of dedifferentiation and differentiation in regenerating amphibian limbs. In *Regeneration,* ed. D. Rudnick, pp. 177–210. New York: Roland Press.

Hay, E. D. (1966). *Regeneration.* New York: Holt, Rinehart, and Winston.

Hay, E. D. (1971). Skeletal muscle regeneration. *New Engl. J. Med. 284:* 1033–4.

Hayamizu, T. F., Wanek, N., Taylor, G., Trevino, C., Shi, C., Anderson, R.,

Gardiner, D. M., Muneoka, K., and Bryant, S. V. (1994). Regeneration of Hox D expression domains during pattern regulation in chick wing buds. *Dev. Biol. 161:* 504–12.

Hearson, L. L., Eltinge, E. M., and Vanable, J.W., Jr. (1988). Do stump fields promote the nerve growth that is necessary for limb regeneration? In *Proceedings of the 6th M. Singer Symposium,* eds. S. Inoue et al., pp. 211–20. Maebashi, Japan: Okada Printing and Publishing Co.

Henderson, C. E., Camu, W., Mettling, C., Couin, A., Poulsen, K., Karihalo, M., Rullamas, J., Evans, T., McMahon, S. B., Armanini, M. P., Beckemeier, L., Phillips, H. S., and Rosenthal, A. (1993). Neurotrophins promote motor neuron survival and are present in embryonic limb bud. *Nature 363:* 266–70.

Herbst, K. (1901). Ueber die regeneration von antenneahulichen organen stelle von augen. *Arch. Entwmech. Org. 13:* 436–47.

Hertwig, G. (1925). Die verpflanzung haploidkerniger zellen, eine neue Methode embryonaler Transplantation. *Roux's Arch. 105:* 294–301.

Hertwig, G. (1927). Beitrage zum determinations und regenerationsproblem mittels der transplantation haploidkerniger zellen. *Roux's Arch. 111:* 292–316.

Hill, D. S., Ragsdale, C. W., Jr., and Brockes, J. P. (1993). Isoform-specific immunological detection of newt retinoic acid receptor δ 1 in normal and regenerating limbs. *Development 117:* 937–45.

Hinterberger, T. J., and Cameron, J. A. (1983). Muscle and cartilage differentiation in axolotl limb regeneration blastema cultures. *J. Exp. Zool. 226:* 399–407.

Hinterberger, T. J., and Cameron, J. A. (1991). Myoblasts and connective-tissue cells in regenerating amphibian limbs. *Ontogenez 21:* 341–57.

Hofmann, C., and Eichele, G. (1994). Retinoids in development. In *The Retinoids: Biology, Chemistry and Medicine,* eds. M. B. Sporn, A. B. Roberts, and D. S. Goodman, pp. 387–441. New York: Raven Press.

Holder, N., and Tank, P. W. (1979). Morphogenetic interactions occurring between blastemas and stumps after exchanging blastemas between normal and double-half forelimbs in the axolotl, *Amblystoma mexicanum. Dev. Biol. 68:* 271–9.

Holder, N. (1981). Pattern formation and growth in the regenerating limbs of urodelean amphibians. *J. Embryol. Exp. Morphol. 65:* 19–36.

Holder, N., and Weekes, C. (1984). Regeneration of surgically created mixed-handed axolotl forelimbs: pattern formation in the dorsal–ventral axis. *J. Embryol. Exp. Morph. 82:* 217–39.

Holtfreter, J. (1955). Transformation of a tail into a limb or gill-like structures. *J. Exp. Zool. 129:* 623–48.

Hondermarck, H., and Boilly, B. (1990). Characterization of fibroblast growth factor binding in regenerating limb blastemas of axolotls. Presented at the European Conference on Tissue and Post-traumatic Regeneration, Geneva,. Sept. 3–7.

Humphrey, R. R. (1967). Genetic and experimental studies on a lethal trait ("Short Toes") in the Mexican axolotl (*Amblystoma mexicanum*). *J. Exp. Zool. 164:* 281–96.

Ide, H., Takamatu, K., and Koshida, K. (1994). Limb bud cells and blastema cells of Xenopus laevis in culture. Presented at the 33rd NIBB Conference on Approaches to the Cellular and Molecular Mechanisms of Regeneration, Okazaki, Japan, March 23–25.

Illingworth, C. M. (1974). Trapped fingers and amputated finger tips in children. *J. Pediatr. Surg. 9:* 853–8.

Imokawa, Y., and Eguchi, G. (1992). Expression and distribution of regeneration-responsive molecule during normal development of the newt, *Cynops pyrrhogaster. Int. J. Dev. Biol. 36:* 407–12.

Imokawa, Y., Ono, S.-I., Takeuchi, T., and Eguchi, G. (1992). Analysis of a unique molecule responsible for regeneration and stabilization of differentiated state of tissue cells. *Int. J. Dev. Biol. 36:* 399–405.

Iten, L. E., and Bryant, S. V. (1973). Forelimb regeneration from different levels of amputation in the newt, *Notophthalmus viridescens*: Length, rate and stages. *Roux's Arch. Entwicklungsmech. Org. 173:* 263–82.

Iten, L. E., and Bryant, S. V. (1975). The interaction between the blastema and stump in the establishment of the anterior–posterior and proximal–distal organization of the limb regenerate. *Dev. Biol. 44:* 119–47.

Izpisúa-Belmonte, J. C.,Tickle, C., Dolle, P., Wolpert, L., and Duboule, D. (1991). Expression of the homeobox Hox-4 genes and the specification of position in chick wing development. *Nature 350:* 585–9.

Jabaily, J. A., Rall, T. W., and Singer, M. (1975). Assay of cyclic 3'-5'-monophosphate in the regenerating forelimb of the newt, *Triturus. J. Morphol. 147:* 379–84.

Jabaily, J. A., Blue, P., and Singer, M. (1982). The culturing of dissociated newt forelimb regenerate cells. *J. Exp. Zool. 219:* 67–73.

Johnson, K. J., and Scadding, S. R. (1992a). Effects of tunicamycin on retinoic acid induced respecification of positional values in regenerating limbs of the larval axolotl, *Ambystoma mexicanum. Develop. Dyn. 193:* 185–92.

Johnson, K. J., and Scadding, S. R. (1992b). The duration of the effectiveness of Vitamin A at causing proximodistal duplication in regenerating limbs of the axolotl, *Ambystoma mexicanum*, in relation to whole body retinoid levels. *J. Exp. Zool.* 264: 189–95.

Johnson, K. J., and Langille, R. M. (1993). The effects of locally applied tunicamycin on anterior–posterior patterning in the developing chick limb. Presented at the 33rd Regional Developmental Biology Conference, Dayton, OH, May 12–14.

Johnson, R. L., and Tabin, C. (1995). The long and short of hedgehog signaling. *Cell 81:* 313–16.

Jordan, M. (1960). Development of regeneration blastemas implanted into the brain. *Folia Biol. (Krakow) 8:* 41–53.

Ju, B.-G., and Kim, W.-S. (1994). Pattern duplication by retinoic acid treatment in the regenerating limbs of the Korean salamander larvae, *Hynobius leechii,* correlates well with the extend of dedifferentiation. *Develop. Dyn. 199.*

Karczmar, A. G. (1946). The role of amputation and nerve resection in the regressing limbs of urodele larvae. *J. Exp. Zool. 103:* 401–27.

Karczmar, A. G., and Berg, G. G. (1951). Alkaline phosphatase during limb development and regeneration of *Amblystoma opacum* and *Amblystoma punctatum. J. Exp. Zool. 117:* 139–63.

Keeble, S., and Maden, M. (1986). Retinoic acid-binding protein in the axolotl: Distribution in mature tissues and time of appearance during limb regeneration. *Dev. Biol. 117:* 435–41.

Kiffmeyer, W. R., Tomusk, E. V., and Mescher, A. L. (1991). Axonal transport and release of transferrin in nerves of regenerating amphibian limbs. *Dev. Biol. 147:* 392–402.

Kim, W.S., and Stocum, D. L. (1986). Retinoic acid modifies positional memory in the anteroposterior axis of regenerating axolotl limbs. *Dev. Biol. 114:* 170–9.

Kintner, C. R., and Brockes J. P. (1984). Monoclonal antibodies identify blastemal cells derived from dedifferentiating muscle in newt limb regeneration. *Nature 508:* 67–9.

Kiortsis, V. (1953). Potentialités du territoire patte chez le Triton (adultes, larves, embryons). *Rev. Suisse Zool. 60:* 301–10.

Kiortsis, V. (1955). La territoire embryonnaire de la patte antérieure du Triton etudié par les greffes hétéroplastiques. *Rev. Suisse Zool. 62:* 171–89.

Kissinger, C. R., Liu, B., Martin-Blanco, E., Kornberg, T. B., and Pabo, C. O. (1990). Crystal structure of an engrailed homeodomain–DNA complex at 2.8 Å resolution. A framework for understanding homeodomain–DNA interactions. *Cell 63:* 579–90.

Kliewer, S. A., Umesono, K., Mangelsdorf, D. J., and Evans, R. M. (1992). Retinoic X receptor interacts with nuclear receptors in retinoic acid, thyroid hormone and vitamin D_3 signalling. *Nature 355:* 446–9.

Kollros, J. J. (1984). Limb regeneration in anuran tadpoles following repeated amputations. *J. Exp. Zool. 232:* 217–29.

Korschelt, E. (1931). Reprinted in *Regeneration and Transplantation.* Canton, MA: Science History Publications, 1990.

Lash, J. W. (1955). Studies on wound closure in urodeles. *J. Exp. Zool. 128:* 13–26.

Lash, J. W. (1963). Studies on the ability of embryonic mesonephros explants to form cartilage. *Dev. Biol. 6:* 219–32.

Lassalle, B. (1980). Are the surface potentials necessary for amphibian limb regeneration? *Dev. Biol. 75:* 460–6.

Lassar, A. B., Paterson, B. M., and Weintraub, H. (1986). Transfection of a DNA locus that mediates the conversion of 10T1/2 fibroblasts into myoblasts. *Cell 47:* 649–56.

Laudet, V., and Stehelin, D. (1992). Flexible friends. *Curr. Biol. 2:* 293–5.

Laufer, E., Nelson, C. E., Johnson, R. L., Morgan, B. A., and Tabin, C. (1994). *Sonic hedgehog* and *Fgf-4* act through a signaling cascade and feedback loop to integrate growth and patterning of the developing limb bud. *Cell 79:* 993–1003.

Laz, T. M., and Sicard, R. E. (1982). Cyclic nucleotide metabolism during amphibian forelimb regeneration. I. The protein kinases. *Roux's Arch. Dev. Biol. 191:* 163–8.

Lazard, L. (1967). Restauration de la régénération de membres irradiés d'axolotl par les greffes hétérotopiques d'origines diverses. *J. Embryol. Exp. Morphol. 18:* 321–42.

Lebowitz, P., and Singer, M. (1970). Neurotrophic control of protein synthesis in the regenerating limb of the newt, *Triturus. Nature 225:* 824–7.

Lechleiter, J., Girard, S., Peralta, E., and Clapham, D. (1991). Spiral calcium wave propagation and annihilation in *Xenopus laevis* oocytes. *Science 252:* 123–6.

Lee, M. S., Kliewer, S. A., Provencal, J., Wright, P. E., and Evans, R. M. (1993). Structure of the retinoic X receptor α DNA binding domain: A helix required for homodimeric DNA binding. *Science 260:* 1117–21.

Lehmann, F. E. (1961). Action of morphostatic substances and the role of proteases in regenerating tissues and in tumor cells. *Adv. Morphog. 1:* 153–87.

Lemaitre, J.-M., Mechali, M., and Geraudie, J. (1992). Nerve-dependent expression of c-myc protein during forelimb regeneration of *Xenopus laevis* froglets. *Int. J. Dev. Biol. 36:* 483–9.

Lemanski, L. F., LaFrance, S., Dube, D. K., Nakatsugawa, M., Erginel-Unaltuna, N., Fransen, M. E., Capone, R. and Lemanski, S. F. (1993). Studies on an RNA which promotes myofibril formation in mutant axolotl hearts. Presented in the International Workshop on the Molecular Biology of Axolotls and Other Urodels, Indianapolis, IN, October 13–16.

Levin, A. A., Sturzenbecker, L. J., Kazmer, S., Bosakowski, T., Huselton, C., Allenby, G., Speck, J., Kratzeisen, C. L., Rosenberger, M., Lovey, A., and Grippo, J. F. (1992). 9-*cis*-retinoic acid stereoisomer binds and activates the nuclear RXRα. *Nature 355:* 359–61.

Lheureux, E. (1975). Régénération des membres irradiés de *Pleurodeles waltlii* Michah. (Urolele). Influence des qualités et orientation des greffons non irradiés. *Roux's Arch. Dev. Biol. 176:* 303–27.

Liversage, R. A., and Colley, B. M. (1965). Effects of puromycin on the differentiation of the regenerating forelimb in *Diemyctulus viridescens. Am. Zool. 5:* 720.

Liversage, R. A., and Globus, M. (1977). *In vitro* regulation of innervated forelimb deplants of *Amblystoma* larvae. *Can. J. Zool. 55:* 1195–9.

Liversage, R. A., Rathbone, M. P., and McLaughlin, H. M. G. (1977). Changes in cyclic GMP levels during forelimb regeneration in adult *Notophthalmus viridescens. J. Exp. Zool. 200:* 169–75.

Liversage, R. A., McLaughlin D. S., and McLaughlin, H. M. G. (1985). The hormonal milieu in amphibian appendage regeneration. In *Regulation of Vertebrate Limb Regeneration,* ed. R. E. Sicard, pp. 54–80. New York: Oxford University Press.

Lo, D. C., Allen, F., and Brockes, J. P. (1993). Reversal of muscle differentiation during urodele limb regeneration. *Proc. Natl. Acad. Sci. USA* 90: 7230–4.

Locatelli, P. (1924). L'influenza del sistema nervoso sui processi di regenerazione. *Arch. Sci. Biol. 5:* 362–78.

Locatelli, P. (1929). Der einfluss des nervensystems auf die regeneration. *Arch. Entwmech. Org. 114:* 686–770.

Ludolph, D. C., Cameron, J. A., and Stocum, D. L. (1990). The effect of retinoic acid on positional memory in the dorso-ventral axis of regenerating axolotl limbs. *Dev. Biol. 140:* 41–52.

Ludolph, D. C., Cameron, J. A., Neff, A. W., and Stocum, D. L. (1993). Cloning and tissue specific expression of the axolotl cellular retinoic acid binding protein. *Dev. Growth Differ. 35:* (3), 341–7.

Maas, R., Elfering, S., Glaser, T., and Jepeal, L. (1994). Deficient outgrowth of the ureteric bud underlies the renal agenesis phenotype in the mice manifesting the *limb deformity (ld)* mutation. *Dev. Dyn. 199:* 214–28.

MacDonald, P. N., Dowd, D. R., Nakajima, S., Galligan, M. A., Reeder, M. C., Haussler, C. A., Ozato, K., and Haussler, M. R. (1993). Retinoid X receptors stimulate and 9-*cis* retinoic acid inhibits 1,25-dihydroxyvitamin D_3-activated expression of the rat osteocalcin gene. *Mol. and Cell. Biol., 13:* 5907–17.

Maden, M., and Wallace, H. (1975). The origin of limb regenerates from cartilage grafts. *Acta Embryol. Exp.:* 77–86.

Maden, M. (1977a). The role of Schwann cells in paradoxical regeneration in the axolotl. *J. Embryol. Exp. Morphol. 41:* 1–13.

Maden, M. (1977b). The regeneration of positional information in the amphibian limb. *J. Theor. Biol. 69:* 735–53.

Maden, M. (1978). Neurotrophic control of cell cycle during amphibian limb regeneration. *J. Embryol. Exp. Morphol. 48:* 169–75.

Maden, M., and Turner, R. N. (1978). Supernumerary limbs in the axolotl. *Nature 273:* 232–5.

Maden, M. (1980). Structure of supernumerary limbs. *Nature 286:* 803–5.

Maden, M. (1981a). Morphallaxis in an epimorphic system: Size, growth control and pattern formation during amphibian limb regeneration. *J. Embryol. Exp. Morphol. 65:* 151–67.

Maden, M. (1981b). Experiments on anuran limb buds and their significance for principles of vertebrate limb development. *J. Embryol. Exp. Morphol. 63:* 243–65.

Maden, M. (1982). Vitamin A and pattern formation in the regenerating limb. *Nature 295:* 672–5.Maden, M., and Mustafa, K. (1982). The structure of 180° supernumerary limbs and a hypothesis of their formation. *Dev. Biol. 93:* 257–65.

Maden, M., Keeble, S., and Cox, R. A. (1985). The characteristics of local application of retinoic acid to the regenerating axolotl limb. *Roux's Arch. Dev. Biol. 194:* 228–35.

Maden, M., and Keeble, S. (1987). The role of cartilage and fibronectin during respecification of pattern induced in the regenerating amphibian limb by retinoic acid. *Differentiation 36:* 175–84.

Maden, M., Ong, D. E., Summerbell, D., and Chytil, F. (1988). Spatial distribution of cellular protein binding to retinoic acid in the chick limb bud. *Nature 335:* 733–5.

Maden, M., Darmon D., and Erikson, U. (1990). Studies on the mechanisms of respicification of positional information by retinoic acid in the regeneration blastema. Presented at the European Conference on Tissue and Post-traumatic Regeneration, Geneva, Sept. 3–7.

Maden, M., Summerbell, D., Maignan, J., Darmon, M., and Shroot, B.
(1991). The respecification of limb pattern by new synthetic retinoids
and their interaction with cellular retinoic acid-binding protein.
Differentiation 47: 49–55.

Maden, M. (1993). The homeotic transformation of tails into limbs in *Rana
temporaria* by retinoids. *Dev. Biol. 159:* 379–91.

Maden, M. (1994). The limb bud – Part two. *Nature 371:* 560–1.

Maier, C. E., Watanabe, M., Singer, M., McQuarrie, I. G., Sunshine, J., and
Rutishauser, U. (1986). Expression and function of neural cell adhesion
molecule during limb regeneration. *Proc. Natl. Acad. Sci. USA 83:*
8395–9.

Maier, C. E., and Miller, R. H. (1992). *In vitro* and *in vivo* characterization of
blastemal cells from regenerating newt limbs. *J. Exp. Zool. 262:* 180–92.

Mailman, M. L., and Dresden, M. H. (1976). Collagen metabolism in the
regenerating forelimb of *Notophthalmus viridescens*: Synthesis, accumulation and maturation. *Dev. Biol. 50:* 378–94.

Marsh, J. L., and Theisen, H. (1993). Molecular requirements of positional
value. Presented at the 12th Marcus Singer Symposium, Irvine, CA,
December 3–4.

Martiel, J. L., and Goldberger, A. (1985). Autonomous chaotic behaviors of
the slime mold *Dictyostelium discoideum* predicted by a model for
cAMP signalling. *Nature 313:* 590–2.

McGinnis, M. E., and Vanable, J. W., Jr. (1986a). Wound epithelium resistance controls stump currents. *Dev. Biol. 116:* 174–83.

McGinnis, M. E., and Vanable, J. W., Jr. (1986b). Electrical fields in
Notophthalmus viridescens limb stumps. *Dev. Biol. 116:* 184–93.

Meinhardt, H. (1993). A bootstrap model for the proximodistal pattern formation in vertebrate limbs. *J. Embryol. Exp. Morphol. 76:* 139–46.

Meredith, J.,Jr., Takada, Y., Fornaro, M., Languino, L. R., and Schwartz,
M. A. (1995). Inhibition of cell cycle progression by the alternatively
spliced integrin beta1c. *Science 269:* 1570–2.

Mescher, A. L., and Tassava, R. A. (1975). Denervation effects on DNA replication and mitosis during the initiation of limb regeneration in adult
newts. *Dev. Biol. 44:* 187–97.

Mescher, A. L. (1976). Effects on adult newt limb regeneration of partial and
complete skin flaps over the amputation surface. *J. Exp. Zool. 195:* 117–28.

Mescher, A. L., and Gospodarowicz, D. (1979). Mitogenic effect of a growth
factor derived from myelin on denervated regenerates of newt forelimbs.
J. Exp. Zool. 207: 497–510.

Mescher, A.L. (1983). Growth factors from nerves and their roles during limb
regeneration. In *Limb Development and Regeneration A,* eds. J. F. Fallon
and A. I. Caplan, pp. 501–12. New York: Alan R. Liss.

Mescher, A. L., and Munaim, S. I. (1984). "Trophic" effect of transferrin on
amphibian limb regeneration blastemas. *J. Exp. Zool. 230:* 485–90.

Mescher, A. L. (1993). Development and regeneration of limbs in the Short
toes axolotl mutant. In *Limb Development and Regeneration,* eds. J. F.
Fallon et al., pp. 181–91. New York: Wiley-Liss.

Mettetal, C. (1952). Action du support sur la différenciation des segments proximaux dans les régénérats de membre des amphibiens urodeles. *CRS Acad. Sci.III 234:* 675.

Michael, M. I., and Faber, J. (1961). The self-differentiation of the paddle-shaped limb regenerate, transplanted with normal and reversed proximal–distal orientation after removal of the digital plate (*Amblystoma mexicanum*). *Arch. Biol. 72:* 301–30.

Michael, M. I., and Aziz, F. K. (1975). Effects of mechanical means on the restoration of the limb regenerative ability in metamorphic stages of *Bufo regularis* reuss. *Acta Biol. Acad. Sci. Hung. 26:* 15–21.

Michael, M. I., Aziz, F. K., and Fahmy, G. H. (1993). Effects of cyclophosphamide on limb regeneration in stages of *Bufo regularis* reuss. In *Limb Development and Regeneration,* eds. J. F. Fallon et al., pp. 213–22. New York: Wiley-Liss.

Milojevic, B. D. (1924). Beiträge zur frage über die determination der regenerate. *Roux's Arch. Entwicklungsmech. Org. 103:* 80–94.

Milojevic, B. D., and Grbic, N. (1925). La régénération et l'inversion de la polarité des extremités chez les tritons adultes à la suite d'une transplantation hétérotope. *C.R. Soc. Biol. (Paris) 93:* 649–51.

Mizell, M. (1968). Limb regeneration: Induction in the newborn opossum. *Science 161:* 283–6.

Mizell, M., and Isaacs, J. J. (1970). Induced regeneration of hindlimbs in the newborn oppossum. *Am. Zool. 10:* 141–55.

Mohanty-Hejmadi, P., Dutta, S. K., and Mahapatra, P. (1992). Limbs generated at site of tail amputation in marbled ballon frog after vitamin A treatment. *Nature 355:* 352–3.

Monroy, A. (1941). Ricerche sulle correnti ellettriche derivabili dalla superficie del corpo di Tritone adulti normali e durante la rigenerazione degli arti e della coda. *Publ. Stat. Zool. Napoli 18:* 265–81.

Morgan, B. A., Izpisúa-Belmonte, J. C., Duboule, D., and Tabin, C. J. (1992). Targeted misexpression of *Hox-4.6* in the avian limb bud causes apparent homeotic transformations. *Nature 358:* 236–9.

Morgan, T. H. (1901). *Regeneration.* London: The Macmillan Company.

Muneoka, K., and Bryant, S. V. (1982). Evidence that the patterning mechanisms in the developing and regenerating limbs are the same. *Nature 298:* 369–71.

Muneoka, K., and Bryant, S. V. (1984a). The cellular contribution to supernumerary limbs in the axolotl *Amblystoma mexicanum. Dev. Biol. 105:* 166–78.

Muneoka, K., and Bryant, S. V. (1984b). Cellular contribution to supernumerary limbs resulting from the interactions between developing and regenerating tissues in the axolotl. *Dev. Biol. 105:* 179–87.

Muneoka, K., Fox, W. F., and Bryant, S. V. (1986). Cellular contribution from dermis and cartilage to regenerating limb blastema in axolotls. *Dev. Biol. 116:* 256–60.

Muneoka, K., Holler-Dinsmore, G. V., and Bryant, S. V. (1986a). Pattern discontinuity, polarity and directional intercalation in axolotl limbs. *J. Embryol. Exp. Morphol. 93:* 51–72.

Muneoka, K., Holler-Dinsmore, G., and Bryant, S. V. (1986b). Intrinsic control of regenerative loss in *Xenopus laevis* limbs. *J. Exp. Zool. 240:* 47–54.

Muneoka, K. (1993). Molecular aspects of limb development and regeneration. Presented at the 12th Marcus Singer Symposium, Irvine, CA, December 3–4.

Namenwirth, M. (1974). The inheritance of cell differentiation during limb regeneration in the axolotl. *Dev. Biol. 41:* 42–56.

Nassonov, N. V. (1936). Influence of various factors on morphogenesis following homotopical subcutaneous insertions of cartilage in the axolotl. *C. R. Dokl. Akad. Sci. URSS 4:* 97–100.

Nassonov, N. V. (1938). Morphogenesis following insertion of the parts of various organs under the skin of the axolotl. *C. R. Dokl. Akad. Sci. URSS 19:* 127–44.

Nassonov, N. V. (1941). Accessory formations developing after the implantation of cartilage under the skin of adult urodele amphibians. *Izdatel. Akad. Nauk SSSR, Moscow* (cited by Carlson, B. M., 1971).

Natnanson, M. A., Hilfer, S. R., and Searles, R. L. (1978). Formation of cartilage by nonchondrogenic cell types. *Dev. Biol. 64:* 99–117.

Needham, A. E. (1952). Regeneration and Wound-Healing. London: Methuen: New York: Wiley.

Neufeld, D. A. (1980). Partial blastema formation after amputation in adult mice. *J. Exp. Zool. 212:* 31–6.

Newcomer, M. E. (1993). Structure of the epididymal retinoic acid binding protein at 2.1 Å resolution. *Structure 1:* 7–18.

Niazi, I. A., and Saxena, S. (1978). Abnormal hind limb regeneration in tadpoles of the toad, *Bufo andersoni*, exposed to excess vitamin A. *Folia Biol. (Krakow) 26:* 3–11.

Nicholas, J. S. (1924a). Regulation of posture in the forelimb of *Amblystoma punctatum. J. Exp. Zool. 40:* 113–59.

Nicholas, J. S. (1924b). Ventral and dorsal implantations of the limb bud in *Amblystoma punctatum. J. Exp. Zool. 39:* 27–41.

Nicholas, J. S. (1926). Extirpation experiments upon the embryonic forelimb of the rat. *Proc. Soc. Exp. Biol. Med. 23:* 436–9.

Nicholas, J. S. (1955). Limb and girdle. In *Analysis of Development,* eds. B. H. Willier, P. Weiss, and V. Hamburger, pp. 429–39. Philadelphia: Saunders.

Nicosia, A., Monaci, P., Tomei, L., De Francesco, R., Nuzzo, M., Stunnenberg, H., and Cortese, R. (1990). A myosin-like dimerization helix and an extralarge homeodomain are essential elements of the tripartite DNA binding structure of LFB1. *Cell 61:* 1225–36.

Niswander, L., Tickle, C., Vogel, A., Booth, I., and Martin, G. R. (1993). FGF-4 replaces the apical ectodermal ridge and directs outgrowth and patterning in the limb. *Cell 75:* 579–87.

Noji, S., Nohno, T., Koyama, E., Muto, K., Ohyama, K., Aoki, Y., Tamura, K., Ohsugi, K., Ide, H., Taniguchi, S., and Saito, T. (1991). Retinoic acid induces polarizing activity but is unlikely to be a morphogen in the chick limb bud. *Nature 350:* 83–6.

Norman, W. P., and Schmidt, A. (1967). The fine structure of tissues in the amputated-regenerating limb of the adult newt *Diemictylus viridescens. J. Morphol. 123:* 271–311.

Oberheim, K. W., and Luther, W. (1958). Versuche uber die extremitatenregeneration von salamanderlarven bei umgekehrter polaritat des amputationsstumpfes. *Roux's Arch. Entwickl. 150:* 373–82.

O'Brien, J. J., and Skinner, D. M. (1988). Characterization of enzymes that degrade crab exoskeleton: II. Two acid proteinase activities. *J. Exp. Zool. 246:* 124–31.

Oka, H. (1934). Zur Analyse experimentell erzeugter Doppelbildungen der extremität-versuche an jungen Larven von Hynobius. *J. Fac. Sci. Tokyo Imperial Univ. 3:* 365–484.

Olmo, E. (1973). Quantitative variations in the nuclear DNA and phylogenesis of the amphibia. *Caryologia 26:* 43–68.

Onda, H., Poulin, M. L., Tassava, R. A., and Chiu, I.-M. (1991). Characterization of a newt tenascin and localization of tenascin mRNA during newt limb regeneration by *in situ* hybridization. *Dev. Biol. 148:* 219–32.

Onda, H., and Tassava, R. A. (1991). Expression of the 9G1 antigen in the apical cap of axolotl regenerates requires nerves and mesenchyme. *J. Exp. Zool. 257:* 336–49.

Orr-Urtreger, A., Bedford, M. T., Burakova, T., Arman, E., Zimmer, Y., Yayon, A., Givol, D., and Lonai, P. (1993). Developmental localization of the splicing alternatives of fibroblast growth factor receptor 2 (FGFR2). *Dev. Biol. 158:* 475–86.

Oudkhir, M., Martelly, I., Castagna, M., Moraczewski, J., and Boilly, B. (1989). Protein kinase C activity during limb regeneration of amphibians. In *Recent Trends in Regeneration Research,* eds. V. Kiortsis, S. Koussoulakos, and H. Wallace, pp. 69–79. New York: Plenum.

Outzen, H. C., Custer, R. P., and Prehn, R. T. (1976). Influence of regenerative capacity and innervation on oncogenesis in the adult frog (*Rana pipiens*). *J. Natl. Cancer Inst. 57:* 79–84.

Papageorgiou, S. (1984). A hierarchical polar coordinate model for epimorphic regeneration. *J. Theor. Biol. 109:* 533–54.

Patrick, J., and Briggs, R. (1964). Fate of cartilage cells in limb regeneration in axolotl (*Amblystoma mexicanum*). *Experientia 20:* 431–2.

Pecorino, L. T., Lo, D. C., and Brockes, J. P. (1994). Isoform-specific induction of a retinoid-responsive antigen after biolistic transfection of chimeric retinoic acid/thyroid hormone receptors into a regenerating limb. *Development 120:* 325–33.

Perez-Castro, A. V., Toth-Rogler, L.E., Wei, L.-N., and Nguyen-Huu, M. C. (1989). Spatial and temporal pattern of expression of the cellular retinoic acid binding protein and the cellular retinol binding protein during mouse embryogenesis. *Proc. Natl. Acad. Sci. USA 86:* 8813–7.

Pescitelli, M. J., Jr. and Stocum, D. L. (1980). The origin of skeletal structures during intercalary regeneration of larval *Ambystoma* limbs. *Dev. Biol. 79:* 255–75.

Polezhaev, L. V. (1937). Über die determination des regenerats einer extremität beim axolotl. *C. R. Dokl. Akad. Sci. URSS 15:* 387–90.

Polezhaev, L. V. (1946). Morphological data on regenerative capacity in tadpole limbs as restored by chemical agents. *C. R. Dokl. Akad. Sci. URSS 54:* 281–4.

Polezhaev, L. V. (1959). Recovery of limb regeneration in axolotls after X-ray irradiation. *DAN SSSR 127:* 630–4.

Polezhaev, L. V. (1966). Mechanisms of recovery of the regenerative capacity suppressed by X-irradiation. *Izvestiya AN SSSR 2:* 254–65.

Polezhaev, L. V. (1972). *Loss and Restoration of Regenerative Capacity in Tissues and Organs of Animals.* Cambridge, MA: Harvard University Press.

Polezhaev, L. V. (1979). Morphogenetic potencies of the regeneration blastema. *Usp. Sovrem. Biol. 87:* 287–303 (In Russian).

Poulin, M. L., Patrie, K. M., Betelho, M. J., Tassava, R. A., and Chiu, I.-M. (1993). Heterogeneity in the expression of fibroblast growth factor receptors during limb regeneration in newts (*Notophthalmus viridescens*). *Development 119:* 353–61.

Poulin, M. L., and Chiu, I.-M. (1995). Re-programming of the expression of KGFR and *bek* variants of fibroblast growth factor receptor 2 during limb regeneration in newts (*Notophthalmus viridescens*). *Dev. Dyn. 202:* 378–87.

Prehn, R. T. (1971). Immunosurveillance, regeneration and oncogenesis. *Progr. Exp. Tumor Res. 14:* 1–24.

Procaccini, D. J., and Doyle, C. M. (1972). The inhibition of limb regeneration in adult *Diemictilus viridescens* treated with streptomycin. *Oncology 26:* 393–404.

Puckett, W. O. (1936). The effects of X-irradiation on limb development and regeneration in *Amblystoma. J. Morphol. 59:* 173–213.

Qian, Y. Q., Billeter, M., Otting, G., Muller, M., Gehring, W. J., and Wuthrich, K. (1989). The structure of the *Antennapedia* homeodomain determined by NMR spectroscopy in solution: Comparison with prokaryotic repressors. *Cell 59:* 573–80.

Ragsdale, C. W., Jr., Petkovich, M., Gates, P. B., Chambon, P., and Brockes, J. P. (1989). Identification of a novel retinoic acid receptor in regenerative tissues of the newt. *Nature 341:* 654–7.

Ragsdale, C. W., Jr., Gates, P. B., and Brockes, J. P. (1992a). Identification and expression pattern of a second isoform of the newt alpha retinoic acid receptor. *Nucleic Acids Res. 20:* 5851.

Ragsdale, C. W., Gates, P. B., Hill, D. S., and Brockes, J. P. (1992b). Delta retinoic acid receptor isoform delta1 is distinguished by its N-terminal sequence and abundance in the limb regeneration blastema. *Mech. Dev. 40:* 99–112.

Rahmani, T., and Kiortsis, V. (1961). Le rôle de la peau et des tissus profonds dans la régénération de la patte. *Rev. Suisse Zool. 68:* 91–102.

Rathbone, M. P., Petri, J., Choo, A. F., Logan, M., Carlone, R. L., and Foret, J. E. (1980). Noradrenaline and cyclic AMP-independent growth stimulation in newt limb blastema. *Nature 283:* 387–8.

Reichmann, E., Schwarz, E. H., Deiner, E. M., Leitner, I., Eilers, M., Berger, J., Busslinger, M., and Beug, H. (1992). Activation of an inducible c-FosER fusion protein causes loss of epithelial polarity and triggers epithelial-fibroblastoid cell conversion. *Cell 71:* 1103–16.

Repesh, L. A., and Oberpriller, J. C. (1978). Scanning electron microscopy of epidermal cell migration in wound healing during limb regeneration in the adult newt *Notophthalmus viridescens. Am. J. Anat. 151:* 539–56.

Repesh, L. A., and Oberpriller, J. C. (1980). Ultrastructural studies of migrating epidermis cells during the wound healing stage of regeneration in the adult newt, *Notophthalmus viridescens. Am. J. Anat. 159:* 187–208.

Richardson, D. (1940). Thyroid and pituitary hormones in relation to regeneration. I. The effects of anterior pituitary hormone on regeneration of the hind leg in normal and thyroidectomised newts. *J. Exp. Zool. 83:* 407–30.

Rose, F. C., Quastler, H., and Rose, S. M. (1955). Regeneration of X-rayed salamander limbs provided with normal epidermis. *Science 122:* 1018–9.

Rose, S. M. (1942). A method for inducing limb regeneration in adult anura. *Proc. Soc. Exp. Biol. Med. 49:* 408–10.

Rose, S. M. (1945). The effects of NaCl in stimulating regeneration of limbs of frogs. *J. Morphol. 77:* 119–39.

Rose, S. M. (1962). Tissue-arc control of regeneration in the amphibian limb. In *Regeneration,* ed. D. Rudnick, pp. 153–209. New York: Roland Press.

Rose, S. M. (1970). *Regeneration: Key to Understanding Normal and Abnormal Growth and Development.* New York: Appleton-Century-Crofts.

Rossant, J., Zirngibl, R., Cado, D., Shago, M., and Giguere, V. (1991). Expression of a retinoic acid response element-hsplacZ transgene defines specific domains of transcriptional activity during mouse embryogenesis. *Genes Dev. 5:* 1333–44.

Ruben, L. N., and Stevens, J. M. (1963). Post-embryonic induction in urodele limbs. *J. Morphol. 112:* 279–301.

Rubin, L., and Saunders, J. W., Jr. (1972). Ectodermal–mesodermal interactions in the growth of limb buds in the chick embryo: Constancy and temporal limits of the ectodermal induction. *Dev. Biol. 28:* 94–112.

Rubin, R. (1903). Versuche uder die Beziehung des Nervensystems zur Regeneration bei Amphibien. *Roux's Arch. Entwmech. Org. 16:* 21–75.

Saunders, J. W., and Gasseling, M. (1968). Ectodermal–mesenchymal interaction in the origin of limb symmetry. In *Epithelial-Mesenchymal Interaction,* eds. R. Fleischmayer and R. E. Billingham, pp. 78–97. Baltimore: Williams and Wilkins.

Savard, P., Gates, P. B., and Brockes, J. P. (1988). Position dependent expression of a homeobox gene transcript in relation to amphibian limb regeneration. *EMBO J. 7:* 4275–82.

Savard, P,. and Tremblay, M. (1995). Differential regulation of Hox C6 in the appendages of adult urodeles and anurans. *J. Mol. Biol. 249:* 879–89.

Scadding, S. R. (1977). Phylogenic distribution of limb regeneration capacity in adult *Amphibia. J. Exp. Zool. 202:* 57–68.

Scadding, S. R., and Vinette, J. L. (1978). A comparison of limb regeneration in diploid and triploid larvae of *Amblystoma (Amphibia, Urodela). Can. J. Zool. 56:* 1715–20.

Scadding, S. R. (1981). Limb regeneration in adult amphibia. *Can. J. Zool. 59:* 34–46.

Scadding, S. R. (1982). Can differences in limb regeneration ability between amphibia species be explained by differences in quantity of innervation? *J. Exp. Zool. 219:* 81–5.

Scadding, S. R., and Maden, M. (1986). Comparison of the effects of vitamin A on development and regeneration in the axolotl, *Amblystoma mexicanum. J. Embryol. Exp. Morphol. 91:* 19–34.

Scadding, S. R., and Maden, M. (1994). Retinoic acid gradients during limb regeneration. *Dev. Biol. 162:* 608–17

Scharf, A. (1961). Experiments on regenerating rat digits. *Growth 25:* 7–23.

Scharf, A. (1963). Reorganization of cornified nail-like outgrowths related with the wound healing process of the amputation sites of the adult newt *Triturus. Growth 27:* 255–69.

Schaxel, J. (1922). Über die Natur der Formvorgänge in der tierischen Entwicklung. *Roux's Arch. Entwicklungsmech. Org. 50:* 498–525.

Schaxel, J (1934). Zur Determination der Regeneration der Axolotl-Extremität. *C. R. Dokl. Akad. Sci. URSS 4:* 246–8.

Schilthuis, J. G., Gann, A., and Brockes, J. P. (1993). Chimeric retinoic acid/thyroid hormone receptors implicate RAR-α1 as mediating growth inhibition by retinoic acid. *EMBO J. 12:* 3459–66.

Schmidt, A. J., and Weary, M. (1963). Localization of acid phosphatase in the regenerating forelimb of the adult newt *Diemictylus viridescens. J. Exp. Zool. 152:* 101–13.

Schmidt, A. J. (1966). *The Molecular Basis of Regeneration: Enzymes.* (Illinois Monographs in Medical Sciences.) Urbana: University of Illinois Press.

Schmidt, A. J. (1968). *Cellular Biology of Vertebrate Regeneration and Repair.* University of Chicago Press.

Schneider, J. W., Gu, W., Zhu, L., Mahdavi, V., and Nadal-Ginard, B. (1994). Reversal of terminal differentiation mediated by p107 in Rb$^{-/-}$ muscle cells. *Science 264:* 1467–71.

Schotte, O. E. (1922). Influences des nerfs sur la régénération des pattes antérieures de tritons adultes. *C. R. Soc. Phys. Hist. Natl., Genève 39:* 67–70.

Schotte, O. E. (1923). Le grand sympathique-élément essentiel de l'influence de système nerveux sur la régénération des pattes de tritons. *C. R. Soc. Phys. Hist. Natl., Genève 39:* 137–9.

Schotte, O. E. (1926). Système nerveux et régénération chez le Triton. *Rev. Suisse Zool. 33:* 1–211.

Schotte, O. E., and Hummel, K. P. (1939). Lens induction at the expense of regenerating tissues of amphibia. *J. Exp. Zool. 80:* 131–66.

Schotte, O. E., and Butler, E. G. (1941). Morphological changes in amputated nerveless limbs of urodele larvae. *Science 93:* 439.

Schotte, O. E., and Harland, M. (1943). Effects of blastema transplantations on regeneration processes of limbs in *Amblystoma* larvae. *Anat. Rec. 87:* 165–80.

Schotte, O. E., and Hall, A. B. (1952). Effects of hypophysectomy upon phases of regeneration in progress (*Triturus viridescens*). *J. Exp. Zool. 121:* 521–60.

Schultz, E. A. (1907). Uber reduktionen. III. Die Reduktion und Regeneration des abgeschnittenen Kiemenkorbes von *Clavellina lepadiformis. Arch. Entw. Meck. 24:* 503.

Schwidefsky, G. (1934). Entwicklung und Determination der extremitaten-regenerate bei den Molchen. *Roux's Arch. Entwicklungsmech. Org. 132:* 57–114.

Scotet, E., Reichmann, E., Breathnach, R., and Houssaint, E. (1995). Oncoprotein Fos activation in epithelial cells induces an epithelio–mesenchymal conversion and changes the receptor encoded by the FGFR-2 mRNA from K-SAM to BEK. *Oncol. Rep. 2:* 203–7.

Sessions, S. K., and Bryant, S. V. (1988). Regenerative ability is an intrinsic property of limb cells in *Xenopus. J. Exp. Zool. 247:* 39–44.

Sherman, D. R., Lloyd, S., and Chytil, F. (1987). Rat cellular retinol-binding protein: cDNA sequence and rapid retinol-dependent accumulation of mRNA. *Proc. Natl. Acad. Sci. USA 84:* 3209–13.

Sicard, R. E. (1989). Epimorphic regeneration and the immune system. In *Recent Trends in Regeneration Research,* eds. V. Kiortsis, S. Koussoulakos, and H. Wallace, pp. 107–119. New York: Plenum.

Sicard, R. E. (1993). Phosphoprotein phosphatase activity in regenerating forelimbs adult newts, *Notophthalmus viridescens.* In *Limb Development and Regeneration,* eds. J. F. Fallon et al., pp. 223–31. New York: Wiley-Liss.

Simeone, A., Acampora, D., Nigro, V., Faiella, A., D'Esposito, M., Stornaiuolo, A., Mavilio, F., and Boncinelli, E. (1991). Differential regulation by retinoic acid of the homeobox genes of the four HOX loci in human embryonal carcinoma cells. *Mech. Dev. 33:* 215–28.

Simon, H.-G., and Tabin, C. J. (1993). Analysis of Hox-4.5 and Hox-3.6 expression during newt limb regeneration: Differential regulation of paralogous Hox genes suggests different roles for members of different Hox clusters. *Development 117:* 1397–1407.

Simon, H.-G., Oppenheimer, S., Nelson, C., Goff, D., and Tabin, C. (1994). Genes controlling amphibian limb regeneration. Presented at the 33rd NIBB Conference on Approaches to the Cellular and Molecular Mechanisms of Regeneration, Okazaki, Japan, March 23–25.

Singer, M. (1942a). The nervous system and regeneration of the forelimb of adult *Triturus.* I: The role of the sympathetics. *J. Exp. Zool. 90:* 37–399.

Singer, M. (1942b). The sympathetics of the brachial region of the urodele, *Triturus. J. Comp. Neurol. 76:* 119–43.

Singer, M. (1943). The nervous system and regeneration of the forelimb of adult *Triturus.* II: The role of the sensory supply. *J. Exp. Zool. 92:* 297–315.

Singer, M. (1945). The nervous system and regeneration of the forelimb of adult *Triturus.* III: The role of the motor supply, including a note on the anatomy of the brachial spinal nerve roots. *J. Exp. Zool. 98:* 1–21.

Singer, M. (1946a). The nervous system and regeneration of the forelimb of adult *Triturus.* IV: The stimulated action of a regenerated motor supply. *J. Exp. Zool. 101:* 221–39.

Singer, M. (1946b). The nervous system and regeneration of the forelimb of adult *Triturus.* V: The influence of number of nerve fibers, including a quantitative study of limb innervation. *J. Exp. Zool. 101:* 299–337.

Singer, M. (1947a). The nervous system and regeneration of the forelimb of the adult *Triturus.* VI: A further study of the importance of nerve number, including quantitative measurements of limb innervation. *J. Exp. Zool. 104:* 223–50.

Singer, M. (1947b). The nervous system and regeneration of the forelimb of the adult *Triturus*. VII: The relation between number of nerve fiber and surface area of amputation. *J. Exp. Zool. 104:* 251–65.

Singer, M. (1951). Induction of regeneration of forelimb of the frog by augmentation of the nerve supply. *Proc. Soc. Exp. Biol. Med. 76:* 413–6.

Singer, M. (1952). The influence of the nerve in regeneration of the amphibian extremity. *Quart. Rev. Biol. 27:* 169–200.

Singer, M. (1954) Induction of regeneration of the forelimb of the postmetamorphic frog by augmentation of the nerve supply. *J. Exp. Zool. 126:* 419–71.

Singer, M. (1961). Induction of regeneration of body parts in the lizard, *Anolis. Proc. Soc. Exp. Biol. Med. 107:* 106–8.

Singer, M., and Salpeter, M. M. (1961). Regeneration in vertebrates. The role of the wound epithelium. In *Growth in Living Systems,* ed. M. X. Zarrow, pp. 277–311. New York: Basic Books.

Singer, M., and Inoue, S. (1964). The nerve and the epidermal apical cap in regeneration of the forelimb of adult *Triturus. J. Exp. Zool. 155:* 105–16.

Singer, M., Maier, C. E., and McNutt, W. S. (1976). Neurotrophic activity of brain extracts in forelimb regeneration of the urodele, *Triturus. J. Exp. Zool. 196:* 131–50.

Singer, M. (1978). On the nature of the neurotrophic phenomenon in urodele limb regeneration. *Am. Zool. 18:* 829–41.

Sisken, B. F., Fowler, I., and Barr, E. (1986). Sources of trophic factors that induce limb regeneration and prevent amputation-induced neuronal death. *Dev. Brain Res. 27:* 181–9.

Skinner, D. M., and Graham, D. E. (1970). Molting in land crabs: Stimulation by leg removal. *Science 169:* 383–5.

Skinner, D. M., and Cook, J. S. (1991). New limbs for old: Some highlights in the history of regeneration in Crustacea. In *A History of Regeneration Research,* ed. C. E. Dinsmore, pp. 25–45. Cambridge University Press.

Skowron, S., and Komala, Z. (1957). Regeneracja konczyn u Xenopus laevis po przeobrazeniu. *Folia Biol. 5:* 53–72.

Skowron, S., and Roguski, H. (1958). Regeneration from implanted dissociated cells. I. Regenerative potentialities of limb and tail cells. *Folia Biol. 6:* 163–73.

Slack, J. M. W. (1976). Determination of polarity in the amphibian limb. *Nature 261:* 44–6.

Slack, J. M. W., and Savage, S. (1978a). Regeneration of reduplicated limbs in contravention of the complete circle rule. *Nature 271:* 760–1.

Slack, J. M. W., and Savage, S. (1978b). Regeneration of mirror-symmetrical limbs in the axolotl. *Cell 14:* 3–14.

Slack, J. M. W. (1980). A serial threshold theory of regeneration. *J. Theor. Biol. 82:* 105–40.

Slack, J. M. W. (1982). Protein synthesis during limb regeneration in the axolotl. *J. Embryol. Exp. Morphol. 70:* 241–60.

Smith, A. R., Lewis, J. H., Crawley, A., and Wolpert, L. (1974). A quantitative study of blastemal growth and bone regression during limb regeneration in *Triturus cristatus. J. Embryol. Exp. Morphol. 32:* 375–90.

Smith, S. D., and Crawford, G. L. (1969). Initiation of regeneration in adult *Rana pipiens* limbs by injection of homologous liver nuclear RNP. *Oncology 23:* 299.

Smith S. D. (1974). Effects of electrode placement on stimulation of adult frog limb regeneration. *Ann. NY Acad. Sci. 238:* 500–7.

Smith, G. N. , Jr., Toole, B. P., and Gross, J. (1975). Hyaluronidase activity and glycosaminoglycan synthesis in the amputated newt limb: Comparison of denervated, nonregenerating limbs with regenerates. *Dev. Biol. 43:* 221–32.

Sordino, P., van der Hoeven, F., and Duboule, D. (1995). Hox gene expression in teleost fins and the origin of vertebrate digits. *Nature 375:* 678–81.

Spallanzani, L. (1769). *An Essay on Animal Reproductions.* (Translated from the Italian, 1768, by M. Maty.) London: T. Becket.

Steen, T. P. (1968). Stability of chondrocyte differentiation and contribution of muscle to cartilage during limb regeneration in the axolotl (*Siredon mexicanum*). *J. Exp. Zool. 167:* 49–78.

Steen, T. P. (1973). The role of muscle cells in *Xenopus* limb regeneration. *Am. Zool. 13:* 1349–50.

Stintson, B. D. (1963). The response of X-irradiated limbs of adult urodeles to normal tissue grafts. I. Effects of autografts of sixty-day forearm regenerates. *J. Exp. Zool. 153:* 37–52.

Stintson, B. D. (1964a). The response of X-irradiated limbs of adult urodeles to normal tissue grafts. II. Effects of autografts of anterior or posterior halves of sixty-day forearm regenerates. *J. Exp. Zool. 155:* 1–24.

Stintson, B. D. (1964b). The response of X-irradiated limbs of adult urodeles to normal tissue grafts. III. Comparative effects of autografts of complete forearm regenerates and longitudinal half regenerates. *J. Exp. Zool. 156:* 1–18.

Stintson, B. D. (1964c). The response of X-irradiated limbs of adult urodeles to normal tissue grafts. IV. Comparative effects of autografts and homografts of complete forearm regenerates. *J. Exp. Zool. 157:* 159–78.

Stocum, D. L. (1968a). The urodele limb regeneration blastema: A self-organizing system. I. Differentiation *in vitro. Dev. Biol. 18:* 441–56.

Stocum, D. L. (1968b). The urodele limb regeneration blastema: A self-organizing system. II. Morphogenesis and differentiation of autografted whole and fractional blastemas. *Dev. Biol. 18:* 457–80.

Stocum, D. L., and Dearlove, G. E. (1972). Epidermal–mesodermal interaction during morphogenesis of the limb regeneration blastemas in larval salamanders. *J. Exp. Zool. 181:* 49–61.

Stocum, D. L., and Melton, D. A. (1977). Self-organizational capacity of distally transplanted regeneration blastemas in larval salamanders. *J. Exp. Zool. 201:* 451–61

Stocum, D. L. (1978). Regeneration of symmetrical hindlimbs in larval salamanders. *Science 200:* 790–3.

Stocum, D. L. (1980). Intercalary regeneration of symmetrical thighs in the axolotl, *Amblystoma mexicanum. Dev. Biol. 79:* 276–95.

Stocum, D. L. (1981). Distal transformation in regenerating double anterior axolotl limbs. *J. Embryol. Exp. Morphol. 65:* 3–18.

Stocum, D.L. (1982). Determination of axial polarity in the urodele limb regeneration blastema. *J. Embryol. Exp. Morphol. 71:* 193–214.

Stocum, D. L. (1984). The urodele limb regeneration blastema. *Differentiation 27:* 13–28.

Stocum, D. L. (1987). Use of retinoids to analyze the cellular basis of positional memory in regenerating amphibian limbs. *Biochem. Cell Biol. 65:* 750–61.

Stocum, D. L., and Maden, M. (1990). Regenerating limbs. *Meth. Enzymol. 190:* 189–201.

Stocum, D. L. (1991). Limb regeneration: A call to arms (and legs). *Cell 67:* 5–8.

Stornaiuolo, A., Acampora, D., Pannese, M., d'Esposito, D., Morelli, F., Migliaccio, E., Rambaldi, M., Faiella, A., Nigro, V., Simeone, A., and Boncinelli, E. (1990). Human HOX genes are differentially activated by retinoic acid in embryonal carcinoma cells according to their position within the four loci. *Cell Differ. Dev. 31:* 119–27.

Stone, L. S. (1966). The fate of amphibian regenerating blastema implanted into lentectomized eyes. *J. Exp. Zool. 162:* 301–10.

Stringfellow, L. A., and Skinner, D. M. (1988). Molt-cycle correlated patterns of synthesis of integumentary proteins in the land crab *Gecarcinus lateralis. Dev. Biol. 128:* 97–110.

Summerbell, D., Lewis, J. H., and Wolpert, L. (1973). Positional information in chick limb morphogenesis. *Nature 224:* 492–6.

Taban, C. H. (1955). Quelques problèmes de régénération chez les urodeles. *Rev. Suise Zool. 62:* 387–468.

Taban, C. H., Cathieni, M., and Constadinidis, J. (1976). Production of retarded, albino regenerates in newts by alpha-methyl-p-tyrosine. *J. Exp. Zool. 197:* 423–7.

Taban, C. H., Cathieni, M., and Schordeter, M. (1978). Cyclic AMP and noradrenaline sensitivity fluctuations in regenerating newt tissues. *Nature 271:* 470–2.

Tabin, C. J. (1989). Isolation of vertebrate limb-identity genes. *Development 105:* 813–20.

Tabin, C. J., and Laufer, E. (1993). Hox genes and serial homology. *Nature 361:* 692–3.

Tam, Y. K., Vethamany-Globus, S., and Globus, M. (1992). Limb amputation and heat shock induce changes in protein expression in the newt, *Notophthalmus viridescens. J. Exp. Zool. 264:* 64–74.

Tank, P. W. (1978). The occurence of supernumerary limbs following blastemal transplantation in the regenerating forelimb of the axolotl, *Amblystoma mexicanum. Dev. Biol. 62:* 143–61.

Tank, P. W. (1981). Pattern formation following 180° rotation of regeneration blastemas in the axolotl, *Amblystoma mexicanum. J. Exp. Zool. 217:* 377–87.

Tassava, R. A. (1969). Hormonal and nutritional requirements for limb regeneration and survival of adult newts. *J. Exp. Zool. 170:* 33–54.

Tassava, R. A., and Mescher, A. L. (1975). The roles of injury, nerves and the wound epidermis during the intiation of amphibian limb regeneration. *Differentiation 4:* 23–4.

Tassava, R. A., and Lloyd, R. M. (1977). Injury requirement for initiation of regeneration of newt limbs which have whole skin grafts. *Nature* 268: 49–50.

Tassava, R. A., and Olsen, C. L. (1982). Higher vertebrates do not regenerate digits and legs because the wound epidermis is not functional: A hypothesis. *Differentiation 22:* 151–5.

Tassava, R. A., and Olsen, C. L. (1985). Neurotrophic influences on cellular proliferation in urodele limb regeneration: *In vivo* experiments. In *Regulation of Vertebrate Limb Regeneration,* ed. R. E. Sicard, pp. 81–92. New York: Oxford University Press.

Tassava, R. A., Johnson-Wint B., and Gross, J. (1986). Regenerate epithelium and skin glands of the adult newt react to the same monoclonal antibody. *J. Exp. Zool 239:* 229–40.

Tassava, R. A., and Acton, R. D. (1989). Distribution of a wound epithelium antigen in embryonic tissues of newts and salamanders. *Ohio J. Sci. 89:* 12–5.

Tassava, R. A. (1993). Comparison of matrix proteins in blastemas of regenerating limbs and limb buds of embryos. Presented at the 12th Annual Marcus Singer Symposium, Irvine, CA, December 3–4.

Tassava, R. A., Castilla, M., Arsanto, J.-P., and Thouveny, Y. (1993). The wound epithelium of regenerating limbs of *Pleurodeles waltl* and *Notophthalmus viridescens*: Studies with mAbs WE3 and WE4, phalloidin, and DNase I. *J. Exp. Zool. 267:* 180–7.

Taylor, G. P., Anderson, R., Reginelli, A. D., and Muneoka, K. (1994). FGF-2 induces regeneration of the chick limb bud. *Dev. Biol. 163:* 282–4.

Taylor, S. M., and Jones, P. A. (1979). Multiple new phenotypes induced in 10T1/2 and 3T3 cells treated with 5-azacytidine. *Cell 17:* 771–9.

Thaller, C., and Eichele, G. (1987). Identification and spatial distribution of retinoids in the developing chick limb bud. *Nature 327:* 625–8.

Thaller, C., Hofmann, C., and Eichele, G. (1993). 9-*cis*-retinoic acid, a potent inducer of digit pattern duplications in the chick wing bud. *Development 118:* 957–65.

Thoms, S. D., and Fallon, J. F. (1980). Pattern regulation and the origin of extra parts following axial misalignments in the urodele limb bud. *J. Embryol. Exp. Morphol. 60:* 33–55.

Thoms, S. D., and Stocum, D. L. (1984). Retinoic acid-induced pattern duplication in regenerating urodele limbs. *Dev. Biol. 103:* 319–28.

Thornton, C. S. (1938a). The histogenesis of the regenerating forelimb of larval *Amblystoma* after exarticulation of the humerus. *J. Morphol. 62:* 219–35.

Thornton, C. S. (1938b). The histogenesis of muscle in regenerating forelimb of larval *Amblystoma punctatum. J. Morphol. 62:* 17–47.

Thornton, C. S. (1943). The effects of colchicine on limb regeneration in larval *Amblystoma. J. Exp. Zool. 92:* 281–95.

Thornton, C. S. (1949). Beryllium inhibition of regeneration. *J. Morphol. 84:* 459–93.

Thornton, C. S. (1957). The effects of apical cap removal on limb regeneration in *Amblystoma* larvae. *J. Exp. Zool.* 134: 357–82.

Thornton, C. S. (1960). Regeneration of asensory limbs of *Amblystoma* larvae. *Copeia 4:* 371–3.

Thornton, C. S., and Thornton, M. T. (1965). The regeneration of accessory limb parts following epidermal cap transplantation in urodeles. *Experientia 21:* 146–51.

Thornton, C. S., and Thornton, M. T. (1970). Recuperation of regeneration in denervated limbs of *Amblystoma* larvae. *J. Exp. Zool. 173:* 293–301.

Tickle, C., Alberts, B. M., Wolpert, L., and Lee, J. (1982). Local application of retinoic acid to the limb bud mimics the action of the polarizing region. *Nature 296:* 564–5.

Todd, J. T. (1823). On the process of reproduction of the members of the aquatic salamanders. *Quart. J. Sci. Lit. Arts 16:* 84–96.

Toole, B. P,. and Gross, J. (1971). The extracellular matrix of the regenerating newt limb: Synthesis and removal of hyaluronate prior to differentiation. *Dev. Biol. 25:* 57–77.

Trampusch, H. A. L. (1956). The effects of X-rays on regenerative capacity. In *Regeneration in Vertebrates,* ed. C. S. Thornton, pp. 83–98. University of Chicago Press.

Trampusch, H. A. L., and Harrebomee, A. E. (1965). Dedifferentiationa prerequisite of regeneration. In *Proceedings, Regeneration in Animals,* pp. 341–74. Amsterdam: North-Holland.

Trampusch, H. A. L. (1966). Regeneration from interocular grafts. *Arch. Zool. Italiano 51:* 787–828.

Tsonis, A. A., Elsner, J. B., and Tsonis, P. A. (1989). On the dynamics of a forced reaction–diffusion model for biological pattern formation. *Proc. Natl. Acad. Sci. USA 86:* 4938–42.

Tsonis, P. A., and Eguchi, G. (1981). Carcinogens on regeneration. Effects on MNNG and 4NQO on limb regeneration in adult newts. *Differentiation 20:* 52–60.

Tsonis, P. A. (1983). Effects of carcinogens on regenerating and non-regenerating limbs in amphibia (review). *Anticancer Res. 3:* 195–202.

Tsonis, P. A., and Eguchi, G. (1983). Effects of a carcinogen *N*-methyl-*N'*-nitro-*N*-nitrosoguanidine on blastema cells and blastema formation in newt limb regeneration. *Dev. Growth Differ. 25:* 201–10.

Tsonis, P. A., and Eguchi, G. (1985). The regeneration of newt limbs deformed in nature. *Experientia 41:* 918–9.

Tsonis, P. A., and Adamson, E. D. (1986). Specific expression of homoeobox-containing genes during induced differentiation of embryonal carcinoma cells. *Biochem. Biophys. Res. Commun. 137:* 520–7.

Tsonis, P. A. (1987). The nature of positional information. *Trends Biochem. Sci. 12:* 249.

Tsonis, P. A., and Del Rio-Tsonis, K. (1988). Spontaneous neoplasms in amphibia. *Tumor Biol. 9:* 221–4.

Tsonis, P. A., and Goetinck, P. F. (1988). Homology of cellular vitamin A-binding protein with DNA-binding proteins. *Biochem. J. 249:* 933–4.

Tsonis, P. A., English, D., and Mescher, A. L. (1991). Increased content of inositol phosphates in amputated limbs of axolotl larvae, and the effect of beryllium. *J. Exp. Zool. 259:* 252–8.

Tsonis, P. A., Mescher A. L., and Del Rio-Tsonis, K. (1992). Protein synthesis in the newt regenerating limb. *Biochem. J. 281:* 665–8.

Tsonis, P. A., Mescher, A. L., Washabaugh, C., and Del Rio-Tsonis, K. (1992). Gene expression during newt limb regeneration. Monogr. *Dev. Biol. 23:* 131–8.

Tsonis, P. A., Washabaugh, C. H., and Del Rio-Tsonis, K. (1992). Protein synthesis in the brain of newts undergoing limb regeneration. *Int. J. Dev. Biol. 36:* 331–4.

Tsonis, P. A. (1993). A comparative two-dimensional gel protein database of the intact and regenerating newt limb. *Electrophoresis 14:* 148–56.

Tsonis, P. A., Del Rio-Tsonis, K., and Washabaugh, C. H. (1993). Analysis of the mutant axolotl Short toes. In *Limb Development and Regeneration,* eds. J. F. Fallon et al., pp. 171–9. New York: Wiley-Liss.

Tsonis, P. A., Washabaugh, C. H., and Del Rio-Tsonis, K. (1994). Morphogenetic effects of 9-*cis*-retinoic acid on the regenerating limbs of the axolotl. *Roux's Arch. Dev. Biol. 203:* 230–4.

Tsonis, P. A., and Del Rio-Tsonis, K. (1995). Protein separation techniques in the study of tissue regeneration. *J. Chromat. A 698:* 361–7.

Tsonis, P. A., Washabaugh, C. H., and Del Rio-Tsonis, K. (1995). Transdifferentiation as a basis for amphibian limb regeneration. *Sem. Cell Biol. 6:* 127–35.

Tsonis, P. A., Del Rio-Tsonis, K., Wallace, J. L., Burns, J. C., Hofmann, M.-C., Millan, J. L., and Washabaugh, C. H. (1996). Insights in amphibian limb regeneration achieved through the application of cellular and molecular methods. *Int. J. Dev. Biol.* (in press).

Tsonis, P. A., and Sargent, M. (1996). 9-*cis*-retinoic acid antagonizes the stimulatory effect of 1,25 dihydroxyvitamin D_3 on chondrogenesis of chick limb bud mesenchymal cells (submitted).

Turing, A. (1952). The chemical basis of morphogenesis. *Philos. Trans. R. Soc. London. Ser. B 237:* 37–72.

Vanable, J. W., Jr. (1991). A history of bioelectricity in development and regeneration. In *A History of Regeneration Research: Milestones in the Evolution of a Science,* ed. C. E. Dinsmore, pp.151–78. Cambridge University Press.

Vethamany-Globus, S., and Liversage, R. A. (1973). Effects of insulin insufficiency on forelimb and tail regeneration in adult *Diemictylus viridescens. J. Embryol. Exp. Morphol. 30:* 427–47.

Vethamany-Globus, S., Globus, M., Darch, A., Milton, G., and Tomlinson B. L. (1984). *In vitro* effects of insulin on macromolecular events in newt limb regeneration blastemata. *J. Exp. Zool. 231:* 63–74.

Vincenti, D. M., and Crawford, K. (1993). Retinoic acid and thyroid hormone may function through similar and competitive pathways in regenerating axolotls. Presented at the International Workshop on the Molecular Biology of Axolotls and Other Urodeles, Indianapolis, IN, October 13–16.

Vorontsova, M. A., and Liosner, L. D. (1960). *Asexual Propagation and Regeneration*. New York: Pergamon.

Wagner, G. P., and Misof, B. Y. (1992). Evolutionary modifications of regenerative capability in vertebrates: A comparative study on teleost pectoral fin regeneration. *J. Exp. Zool. 261:* 62–78.

Wallace, H. (1972). The components of regrowing nerves which support the regeneration of irradiated salamander limbs. *J. Embryol. Exp. Morphol. 28:* 419–35.

Wallace, H., and Watson, A. (1979). Duplicated axolotl regenerates. *J. Embryol. Exp. Morphol. 49:* 243–58.

Wallace, H. (1981). *Vertebrate Limb Regeneration*. New York: Wiley.

Walter, F. K. (1910). Schildruse und regeneration. *Arch. Entwmech. 31:* 91–130.

Walter, F. K. (1919). Experimentalle untersuchungen uber die morphogenetische bedeutung des nervensystems. *Anat. Hefle (Arbeiten) 57:* 651–77.

Wanek, N., Muneoka, K., and Bryant, S. V. (1989). Evidence for regulation following amputation and tissue grafting in the developing mouse limb. *J. Exp. Zool. 249:* 55–61.

Wanek, N., Gardiner, D. M., Muneoka, K., and Bryant, S. V. (1991). Conversion by retinoic acid of anterior cells into ZPA cells in the chick wing bud. *Nature 350:* 81–3.

Washabaugh, C. H., Del Rio-Tsonis, K., and Tsonis, P. A. (1993). Variable manifestations in the Short toes (s) mutation of the axolotl. *J. Morphol. 218:* 107–14.

Washabaugh, C. H., and Tsonis, P. A. (1994). Mononuclear leukocytes in the newt limb blastema: *in vitro* behavior. *Int. J. Dev. Biol. 38:* 745–9.

Washabaugh, C. H., and Tsonis, P. A. (1995). Effects of vitamin D on axolotl limb regeneration. *Dev. Growth Diff. 37:* 497–503.

Wei, Y., and Tassava, R. A. (1994). Expression of the $\alpha 1$ chain of the type XII collagen gene during newt forelimb regeneration and its independence on nerve and wound healing. Presented at the 34th Midwest Regional Developmental Biology Conference, Granville, OH, May 23–25.

Weis, J. P., and Weis, J. S. (1970). The effect of nerve growth factor on limb regeneration in *Amblystoma. J. Exp. Zool. 174:* 73–8.

Weis, J. S. (1971). The effects of nerve growth factor on bullfrog tadpoles (*Rana catesbiana*) after limb amputation. *J. Exp. Zool. 180:* 385–92.

Weismann, A. (1892). *Das Keimplasma*. Jena: Gustav Fisher.

Weiss, P. (1922). Abhangigkeit der Regeneration entwickelter amphibienextremitaten von Nervensystem. *Biol. Versuchsnst. Akad. Wiss. Wien Anz. Mitt. 82:* 22–3.

Weiss, P. (1925). Unabhängigkeit der Extremitäten-regeneration von dem Skelett (bei *Triton cristatus*). *Roux's Arch. Entwicklungsmech. Org 104:* 359–94.

Weiss, P. (1927). Potenzprufung am Regenerationsblastem. I. Extremitatenbildung aus schwanzblastem im extremitatenfeld dei Triton. *Roux's Arch. Entwicklungsmech. Org. 111:* 317–40.

Weiss, P. (1930). Potenzprüfung am regenerationsblastem. II. Das Verhalten des Schwanzblastems nach Transplantation an die Stelle der Vorderextremität bei Eidechsen (Lacerta). *Roux's Arch. Entwicklungsmech. Org. 122:* 379–94.

Wheelock, M. J., and Knudsen, K. A. (1991). Cadherins and associated proteins. *In Vivo 5:* 505–13.

Wilkerson, J. A. (1963). The role of growth hormone in regeneration of the forelimb of the hypophysectomized newt. *J. Exp. Zool. 154:* 223–30.

Winfree, A. T. (1987). *When Time Breaks Down.* Princeton University Press.

Wolff, G. (1902). Die physiologische Grundlage der Lehre von den Degenerationszeichen. *Vircow's Arch. 169:* 308–32.

Wolpert, L. (1969). Positional information and the pattern of cellular differentiation. *J. Theor. Biol. 25:* 1–47.

Wolsky, A. (1974). The effect of chemicals with gene-inhibiting activity on regeneration. *Neoplasia Cell Diff.* 153–88.

Yang, E. V., Shima, D. T., and Tassava, R. A. (1992). Monoclonal antibody ST1 identifies an antigen that is abundant in the axolotl and newt limb stump but is absent from the undifferentiated regenerate. *J. Exp. Zool. 264:* 337–50.

Yang, E. V., Huynh, D., and Bryant, S. V. (1993). The 90 x 10^3 Mr procollagenase is upregulated during limb and tail regeneration in *Amblystoma mexicanum.* Presented at the 33rd Annual Midwest Regional Developmental Biology Conference, Dayton, OH, May 12–14.

Yntema, C. L. (1949). Relations between innervation and regeneration of the forelimb in urodele larvae. *Anat. Rec. 103:* 524.

Zeleny, C. (1916). Studies on the factors controlling the rate of regeneration. Illinois Biological Monographs 3. Urbana: University of Illinois.

Zilakos, N. P., Del Rio-Tsonis, K., Tsonis, P. A., and Parchment, R. E. (1992). Newt squamous carcinoma proves phylogenetic conservation of tumors as caricatures of tissue renewal. *Cancer Res. 52:* 4858–65.

Index

actin 28
activin 120
ADP-ribosyl transferase 85
AER 31,186,187
acid phosphatase 34
alkaline phosphatase 36
amiloride 101, 102
amphibia
   capacity of regeneration 9
   gene introns 10
aneurogenic limbs 67
anura
   limb regeneration 9, 94
arthropods 17, 18
autopodium 63
autotomy 11, 17, 18
average model 122
axis
   anterior–posterior 132, 137, 143
   determination 132
   dorsal–ventral 132, 137, 143
   proximal–distal 133, 155, 156
5-azacytidine 86

bek 31
beryllium 19, 20, 21

bipolar cells 87
blastema
   pluripotency 125
   self-organization 125
   transplantation 125, 132
      contralateral 137, 143,
         144
      ipsilateral 145
      proximal–distal 146, 184, 185
blastema cells
   differentiation 36, 37
   division 33
   metabolism 58, 59
   metaplasia 38, 39
   origin 33, 34, 35
   X-irradiation 33
boundary model 120

calcium 74
cancer 58, 60
carcinogens
   and limb regeneration 60
cAMP 71, 73
catecholamines 73
catenins 55, 56
catepsin 58

237